NANOTECHNOLOGY APPLICATIONS IN AGRICULTURAL AND BIOPROCESS ENGINEERING

Farm to Table

Innovations in Agricultural & Biological Engineering

NANOTECHNOLOGY APPLICATIONS IN AGRICULTURAL AND BIOPROCESS ENGINEERING

Farm to Table

Edited by

Megh R. Goyal, PhD, PE
Santosh K. Mishra, PhD
Lohith Kumar Dasarahalli-Huligowda, PhD

AAP | APPLE
ACADEMIC
PRESS

First edition published 2023

Apple Academic Press Inc.
1265 Goldenrod Circle, NE,
Palm Bay, FL 32905 USA

4164 Lakeshore Road, Burlington,
ON, L7L 1A4 Canada

CRC Press
6000 Broken Sound Parkway NW,
Suite 300, Boca Raton, FL 33487-2742 USA

2 Park Square, Milton Park,
Abingdon, Oxon, OX14 4RN UK

© 2023 by Apple Academic Press, Inc.

Apple Academic Press exclusively co-publishes with CRC Press, an imprint of Taylor & Francis Group, LLC

Library and Archives Canada Cataloguing in Publication

Title: Nanotechnology applications in agricultural and bioprocess engineering : farm to table / edited by Megh R. Goyal, PhD, PE, Santosh K. Mishra, PhD, Lohith Kumar Dasarahalli-Huligowda, PhD.
Names: Goyal, Megh R., editor. | Mishra, Santosh K., editor. | Dasarahalli-Huligowda, Lohith Kumar, editor.
Series: Innovations in agricultural and biological engineering.
Description: First edition. | Series statement: Innovations in agricultural and biological engineering | Includes bibliographical references and index.
Identifiers: Canadiana (print) 20210393432 | Canadiana (ebook) 20210393467 | ISBN 9781774637500 (hardcover) | ISBN 9781774637517 (softcover) | ISBN 9781003277439 (ebook)
Subjects: LCSH: Agricultural innovations. | LCSH: Nanotechnology.
Classification: LCC S494.5.I5 N36 2023 | DDC 620/.5—dc23

Library of Congress Cataloging-in-Publication Data

Names: Goyal, Megh R., editor. | Mishra, Santosh K., editor. | Dasarahalli-Huligowda, Lohith Kumar, editor.
Title: Nanotechnology applications in agricultural and bioprocess engineering : farm to table / Megh R. Goyal, Santosh K. Mishra, Lohith Kumar Dasarahalli-Huligowda.
Other titles: Innovations in agricultural and biological engineering.
Description: First edition. | Palm Bay, FL, USA : Apple Academic Press, 2023. | Series: Innovations in agricultural and biological engineering | Includes bibliographical references and index. | Summary: "This new volume looks at the current scenario of research and advances of the use of nanotechnology applications in agricultural and bioprocess engineering. The first section deals with the impact of nanotechnology in agricultural engineering, looking at the role of nanomaterials in plant growth and nutrition. It goes on to discuss specific methods and processes for the application of nanotechnology in the development of food products, nutraceuticals, and therapeutics. This includes nanotechnological methods for iron fortification of dairy food, for processing and preservation of meat and meat products, for selective targeting of cancer, and more. The book goes on to discuss the role of nanotechnology in bioprocessing, such as for biofuel production, for wastewater treatment, and as enzymatic nanoparticles for fabrication processes. Illuminating the various nanotechnological applications for agricultural and bioprocess engineering, Nanotechnology Applications in Agricultural and Bioprocess Engineering: Farm to Table is a rich resource for advanced students, faculty, scientists, researchers, and industry professionals in the field of nanotechnology. This book will be further useful to engineering, pharma, and food quality practitioners who need to become familiar with updated information pertaining to their work"-- Provided by publisher.
Identifiers: LCCN 2021059458 (print) | LCCN 2021059459 (ebook) | ISBN 9781774637500 (hardback) | ISBN 9781774637517 (paperback) | ISBN 9781003277439 (ebook)
Subjects: LCSH: Nanotechnology. | Agricultural innovations. | Agricultural biotechnology.
Classification: LCC S494.5.B563 N336 2023 (print) | LCC S494.5.B563 (ebook) | DDC 338.1/6—dc23/eng/20211221
LC record available at https://lccn.loc.gov/2021059458
LC ebook record available at https://lccn.loc.gov/2021059459

ISBN: 978-1-77463-750-0 (hbk)
ISBN: 978-1-77463-751-7 (pbk)
ISBN: 978-1-00327-743-9 (ebk)

ABOUT THE BOOK SERIES: INNOVATIONS IN AGRICULTURAL AND BIOLOGICAL ENGINEERING

Under this book series, Apple Academic Press Inc. is publishing book volumes over a span of 8–10 years in the specialty areas defined by the American Society of Agricultural and Biological Engineers (www.asabe. org). Apple Academic Press Inc. aims to be a principal source of books in agricultural and biological engineering. We welcome book proposals from readers in areas of their expertise.

The mission of this series is to provide knowledge and techniques for agricultural and biological engineers (ABEs). The book series offers high-quality reference and academic content on agricultural and biological engineering (ABE) that is accessible to academicians, researchers, scientists, university faculty and university-level students, and professionals around the world.

Agricultural and biological engineers ensure that the world has the necessities of life, including safe and plentiful food, clean air and water, renewable fuel and energy, safe working conditions, and a healthy environment by employing knowledge and expertise of the sciences, both pure and applied, and engineering principles. Biological engineering applies engineering practices to problems and opportunities presented by living things and the natural environment in agriculture.

ABE embraces a variety of the following specialty areas (www. asabe.org): aquaculture engineering, biological engineering, energy, farm machinery and power engineering, food, and process engineering, forest engineering, information, and electrical technologies, soil, and water conservation engineering, natural resources engineering, nursery, and greenhouse engineering, safety, and health, and structures and environment.

For this book series, we welcome chapters on the following specialty areas (but not limited to):

1. Academia to industry to end-user loop in agricultural engineering.
2. Agricultural mechanization.
3. Aquaculture engineering.
4. Biological engineering in agriculture.
5. Biotechnology applications in agricultural engineering.

6. Energy source engineering.
7. Farm to fork technologies in agriculture.
8. Food and bioprocess engineering.
9. Forest engineering.
10. GPS and remote sensing potential in agricultural engineering.
11. Hill land agriculture.
12. Human factors in engineering.
13. Impact of global warming and climatic change on agriculture economy.
14. Information and electrical technologies.
15. Irrigation and drainage engineering.
16. Micro-irrigation engineering.
17. Milk Engineering.
18. Nanotechnology applications in agricultural engineering.
19. Natural resources engineering.
20. Nursery and greenhouse engineering.
21. Potential of phytochemicals from agricultural and wild plants for human health.
22. Power systems and machinery design.
23. Robot engineering and drones in agriculture.
24. Rural electrification.
25. Sanitary engineering.
26. Simulation and computer modeling.
27. Smart engineering applications in agriculture.
28. Soil and water engineering.
29. Structures and environment engineering.
30. Waste management and recycling.
31. Any other focus areas.

Books in the Innovations in Agricultural & Biological Engineering Series

- Biological and Chemical Hazards in Food and Food Products: Prevention, Practices, and Management
- Bioremediation and Phytoremediation Technologies in Sustainable Soil Management, 4-volume set:
  - o Volume 1: Fundamental Aspects and Contaminated Sites
  - o Volume 2: Microbial Approaches and Recent Trends
  - o Volume 3: Inventive Techniques, Research Methods, and Case Studies
  - o Volume 4: Degradation of Pesticides and Polychlorinated Biphenyls

- Dairy Engineering: Advanced Technologies and Their Applications
- Developing Technologies in Food Science: Status, Applications, and Challenges
- Emerging Technologies in Agricultural Engineering
- Engineering Interventions in Agricultural Processing
- Engineering Interventions in Foods and Plants
- Engineering Practices for Agricultural Production and Water Conservation: An Interdisciplinary Approach
- Engineering Practices for Management of Soil Salinity: Agricultural, Physiological, and Adaptive Approaches
- Engineering Practices for Milk Products: Dairyceuticals, Novel Technologies, and Quality
- Field Practices for Wastewater Use in Agriculture: Future Trends and Use of Biological Systems
- Flood Assessment: Modeling and Parameterization
- Food Engineering: Emerging Issues, Modeling, and Applications
- Food Process Engineering: Emerging Trends in Research and Their Applications
- Food Processing and Preservation Technology: Advances, Methods, and Applications
- Food Technology: Applied Research and Production Techniques
- Functional Dairy Ingredients and Nutraceuticals: Physicochemical, Technological, and Therapeutic Aspects
- Handbook of Research on Food Processing and Preservation Technologies, 5-volume set:
  o Volume 1: Nonthermal and Innovative Food Processing Methods
  o Volume 2: Nonthermal Food Preservation and Novel Processing Strategies
  o Volume 3: Computer-Aided Food Processing and Quality Evaluation Techniques
  o Volume 4: Design and Development of Specific Foods, Packaging Systems, and Food Safety
  o Volume 5: Emerging Techniques for Food Processing, Quality, and Safety Assurance
- Modeling Methods and Practices in Soil and Water Engineering
- Nanotechnology and Nanomaterial Applications in Food, Health, and Biomedical Sciences
- Nanotechnology Applications in Agricultural and Bioprocess Engineering: Farm to Table

- Nanotechnology Applications in Dairy Science: Packaging, Processing, and Preservation
- Nanotechnology Horizons in Food Process Engineering, 3-volume set:
  - o Volume 1: Food Preservation, Food Packaging and Sustainable Agriculture
  - o Volume 2: Scope, Biomaterials, and Human Health
  - o Volume 3: Trends, Nanomaterials, and Food Delivery
- Novel Dairy Processing Technologies: Techniques, Management, and Energy Conservation
- Novel Processing Methods for Plant-Based Health Foods: Extraction, Encapsulation and Health Benefits of Bioactive Compounds
- Novel Strategies to Improve Shelf-Life and Quality of Foods: Quality, Safety, and Health Aspects
- Processing of Fruits and Vegetables: From Farm to Fork
- Processing Technologies for Milk and Milk Products: Methods, Applications, and Energy Usage
- Quality Control in Fruit and Vegetable Processing: Methods and Strategies
- Scientific and Technical Terms in Bioengineering and Biological Engineering
- Soil and Water Engineering: Principles and Applications of Modeling
- Soil Salinity Management in Agriculture: Technological Advances and Applications
- State-of-the-Art Technologies in Food Science: Human Health, Emerging Issues and Specialty Topics
- Sustainable Biological Systems for Agriculture: Emerging Issues in Nanotechnology, Biofertilizers, Wastewater, and Farm Machines
- Sustainable Nanomaterials for Biosystem Engineering: Impacts, Challenges, and Future Prospects
- Technological Interventions in Dairy Science: Innovative Approaches in Processing, Preservation, and Analysis of Milk Products
- Technological Interventions in Management of Irrigated Agriculture
- Technological Interventions in the Processing of Fruits and Vegetables
- Technological Processes for Marine Foods, from Water to Fork: Bioactive Compounds, Industrial Applications, and Genomics

OTHER BOOKS ON AGRICULTURAL AND BIOLOGICAL ENGINEERING FROM APPLE ACADEMIC PRESS, INC.

Management of Drip/Trickle or Micro Irrigation
Megh R. Goyal, PhD, PE, Senior Editor-in-Chief

Evapotranspiration: Principles and Applications for Water Management
Megh R. Goyal, PhD, PE, and Eric W. Harmsen, Editors

Book Series: Research Advances in Sustainable Micro Irrigation
Senior Editor-in-Chief: Megh R. Goyal, PhD, PE

Volume 1: Sustainable Micro Irrigation: Principles and Practices
Volume 2: Sustainable Practices in Surface and Subsurface Micro Irrigation
Volume 3: Sustainable Micro Irrigation Management for Trees and Vines
Volume 4: Management, Performance, and Applications of Micro Irrigation Systems
Volume 5: Applications of Furrow and Micro Irrigation in Arid and Semi-Arid Regions
Volume 6: Best Management Practices for Drip Irrigated Crops
Volume 7: Closed Circuit Micro Irrigation Design: Theory and Applications
Volume 8: Wastewater Management for Irrigation: Principles and Practices
Volume 9: Water and Fertigation Management in Micro Irrigation
Volume 10: Innovation in Micro Irrigation Technology

Book Series: Innovations and Challenges in Micro Irrigation
Senior Editor-in-Chief: Megh R. Goyal, PhD, PE

Volume 1: Sustainable Micro Irrigation Design Systems for Agricultural Crops
Volume 2: Principles and Management of Clogging in Micro Irrigation
Volume 3: Performance Evaluation of Micro Irrigation Management

ABOUT THE SENIOR EDITOR-IN-CHIEF

Megh R. Goyal, PhD, PE, is a Retired Professor in Agricultural and Biomedical Engineering from the General Engineering Department in the College of Engineering at the University of Puerto Rico–Mayaguez Campus; and Senior Acquisitions Editor and Senior Technical Editor-in-Chief in Agriculture and Biomedical Engineering for Apple Academic Press, Inc. He has worked as a Soil Conservation Inspector and as a Research Assistant at Haryana Agricultural University and Ohio State University.

During his professional career of 52 years, Dr. Goyal has received many prestigious awards and honors. He was the first agricultural engineer to receive the professional license in Agricultural Engineering in 1986 from the College of Engineers and Surveyors of Puerto Rico. In 2005, he was proclaimed as "Father of Irrigation Engineering in Puerto Rico for the Twentieth Century" by the American Society of Agricultural and Biological Engineers (ASABE), Puerto Rico Section, for his pioneering work on micro irrigation, evapotranspiration, agroclimatology, and soil and water engineering. The Water Technology Centre of Tamil Nadu Agricultural University in Coimbatore, India, recognized Dr. Goyal as one of the experts "who rendered meritorious service for the development of micro irrigation sector in India" by bestowing the Award of Outstanding Contribution in Micro Irrigation. This award was presented to Dr. Goyal during the inaugural session of the National Congress on "New Challenges and Advances in Sustainable Micro Irrigation" held at Tamil Nadu Agricultural University.

Dr. Goyal received the Netafim Award for Advancements in Microirrigation: 2018 from the American Society of Agricultural Engineers at the ASABE International Meeting in August 2018. VDGOOD Professional Association of India awarded Lifetime Achievement Award at 12th Annual Meeting on Engineering, Science and Medicine that was held on 20–21 of November of 2020 in Visakhapatnam, India.

A prolific author and editor, he has written more than 200 journal articles and textbooks and has edited over 100 books. He is the editor of three book series published by Apple Academic Press: Innovations in Agricultural & Biological Engineering, Innovations and Challenges in Micro Irrigation, and Research Advances in Sustainable Micro Irrigation. He is also instrumental in the development of the new book series Innovations in Plant Science for Better Health: From Soil to Fork.

Dr. Goyal received his BSc degree in engineering from Punjab Agricultural University, Ludhiana, India; his MSc and PhD degrees from Ohio State University, Columbus; and his Master of Divinity degree from Puerto Rico Evangelical Seminary, Hato Rey, Puerto Rico, USA.

ABOUT THE CO-EDITORS

Santosh Kumar Mishra, PhD, is working as an Assistant Professor in the Department of Dairy Microbiology, College of Dairy Science and Technology, Guru Angad Dev Veterinary and Animal Sciences University, Ludhiana, Punjab, India.

He received his BTech degree in Dairy Technology from Maharashtra Animal and Fisheries Sciences University, Nagpur, India; and his MSc and PhD degrees from the National Dairy Research Institute, Karnal, Haryana. Now he is working in the area of functional foods and dairy products incorporating live probiotics and technology of functional lactic cultures for fermented and non-fermented dairy products. He has also served the dairy industry as Quality Assurance Executive at Mother Dairy, New Delhi.

In India, he has handled externally funded projects granted by DST, MoFPI, and UGC as PI or Co-PI. He received several awards for best papers and posters/presentations. He is the recipient of junior and senior research fellowships during his master's and doctoral programs at the National Dairy Research Institute, Karnal, Haryana. He was a university gold medalist during his graduation program and also earned the best thesis award by his university after his PhD research. Recently he received an Award of Honor at an international conference sponsored by Partap College of Education, Ludhiana, in association with the International Professionals Development Association, UK.

He is a member of various scientific societies: life member of SASNET-Fermented Foods, Anand; member of Indian Dairy Associations, New Delhi. He has published several research, review, and popular articles in national and international journals. He has also published three international edited books, several book chapters, and teaching reviews in various training programs. He has recently completed a young scientist project by DST, SEED Department, Govt. of India, New Delhi, on isolation, and characterization of novel oxalate degrading lactic acid bacteria for potential probiotic management of kidney stone.

Readers may contact him at: *skmishra84@gmail.com.*

Lohith Kumar Dasarahalli-Huligowda, PhD, has obtained his doctorate degree in Bioprocess Engineering at the Department of Biotechnology, Indian Institute of Technology Roorkee, India. He has expertise in nanoemulsion design and formulation. He has worked at CSIR-CFTRI, Mysore, India.

His current professional interests include the utilization of nanotechnology principles to develop delivery systems and biofuels properties enhancement. His main research activities include nano-patterning techniques for food application, fabrication of edible nanostructures, kinetic studies on nano-enabled food matrices structure, functional compound extraction techniques, and utilization of microbiota for bioremediation and biofuel production, and bioprocesses simulation and optimization. He has published scientific articles in international peer-reviewed journals and has authored an edited book and many book chapters as well as review articles. He has published research papers and abstracts mainly on the emulsion-based delivery systems to protect flaxseed oil from oxidation process.

His BTech degree was in Food Science and Technology. With an Indian Council of Agricultural Research Junior Research Fellowship and Ministry of Human Resource Development-GATE, he received an MTech degree in Food Process Engineering from the National Institute of Technology Rourkela, India. His master's research focused on the development of emulsion-based matrices for the protection of flaxseed oil using food grade biopolymers.

Readers can contact him at: *lohithhanum8@gmail.com.*

CONTENTS

CONTRIBUTORS

Chandrajit Balomajumder
Professor, Department of Chemical Engineering, Indian Institute of Technology-Roorkee, Roorkee–247667, Uttarakhand, India, E-mail: chandfch@iitr.ac.in

Soumitra Banerjee
Assistant Professor, Center for Incubation, Innovation, Research, and Consultancy (CIIRC), Jyothy Institute of Technology, Tataguni, off Kanakapura Main Road, Bangalore–560062, Karnataka, India, Mobile: +91-9480443846, E-mail: soumitra.banerjee7@gmail.com

Ajay Kumar Chauhan
PhD Research Scholar, Department of Biotechnology, Indian Institute of Technology, Haridwar Highway, Roorkee–247667, Uttarakhand, India, Mobile +91-7599057306, E-mail: ajaychauhan1408@gmail.com

Lohith Kumar Dasarahalli-Huligowda
Assistant Professor, Department of Biotechnology, Indian Institute of Technology, Haridwar Highway, Roorkee – 247667, Uttarakhand, India, Mobile: +91-7064655392, E-mail: lohithhanum8@gmail.com

Megh R. Goyal
Retired Faculty in Agricultural and Biomedical Engineering from College of Engineering at University of Puerto Rico-Mayaguez Campus; and Senior Technical Editor-in-Chief in Agricultural and Biomedical Engineering for Apple Academic Press Inc., PO Box 86, Rincon-PR–006770086, USA, E-mail: goyalmegh@gmail.com

Gajanan Gundewadi
PhD Research Scholar, Division of Food Science and Postharvest Technology, Indian Agricultural Research Institute (IARI), New Delhi–110012, India, Mobile: +91-7065157693, E-mail: aggajanan21@gmail.com

Nikhil Kumar
PhD Research Scholar, Indian Institute of Technology, Roorkee – 247667, Uttarakhand, India, Mobile: +91-8730852580, E-mail: nk738684@gmail.com

Pawan Kumar
PhD Research Scholar, Dalhousie University, 6299 South St. Halifax, NS B3H 4R2, Canada, Mobile: +91-8699805153, E-mail: pawan.02baps@gmail.com

B. Manjula
Assistant Professor, Department of Agricultural Processing and Food Engineering, College of Agricultural Engineering, Madakasira, Acharya N. G. Ranga Agriculture University, Guntur–515301, Andhra Pradesh, India, Mobile: +91-9703582436, E-mail: manjulaprakash08@gmail.com

Santosh K. Mishra
Assistant Professor, Department of Dairy Microbiology, College of Dairy Science and Technology, Guru Angad Dev Veterinary and Animal Sciences University (GADVASU), Ludhiana–141004, Punjab, India, Mobile: +91-9464995049, E-mail: skmishra84@gmail.com

H. B. Muralidhara
Associate Professor, Center for Incubation, Innovation, Research, and Consultancy (CIIRC), Jyothy Institute of Technology, Tataguni, off Kanakapura Main Road, Bangalore–560062, Karnataka, India, Mobile: +91-9739315239, E-mail: hb.murali@gmail.com

Brijesh Patil Muder Pakeerappa
PhD Research Scholar, Department of Biotechnology, University of Agriculture-Dharwad, Dharwad–580005, Karnataka, India, Mobile: +91-8762185959, E-mail: patil5959@gmail.com

Amrita Poonia
Assistant Professor, Center of Food Science and Technology, Institute of Agricultural Science, Banaras Hindu University, Varanasi–221005, Uttar Pradesh, India, Mobile: +91–9532030058, E-mail: dramritapoonia@gmail.com

Vijay S. Rakesh Reddy
Scientist (Horticulture), ICAR-Central Institute for Arid Horticulture, Sri Ganganagar Road, Beechhwal Rural, Bikane–334006, Rajasthan, India, Mobile: +91-8527034568, E-mail: drrakesh.reddy968@gmail.com

Sahely Saha
PhD Research Scholar, Department of Biotechnology and Medical Engineering, National Institute of Technology Rourkela, Rourkela–769008, Odisha, India, Mobile: +91-9778987894, E-mail: sahelisaha16@gmail.com

Rahul Saini
PhD Research Scholar, Department of Civil Engineering, Lassonde School of Engineering, York University, North York, Ontario, Toronto, M3J 1P3, Canada, Mobile: +91-8699361068, E-mail: sainirahul532@gmail.com

Preetam Sarkar
Assistant Professor, Department of Food Process Engineering, National Institute of Technology, Rourkela – 769008, Orissa, India, Mobile: +91-8971221033, E-mail: sarkarpreetam@nitrkl.ac.in

Nainsi Saxena
PhD Research Scholar, Department of Chemical Engineering, National Institute of Technology Rourkela, Rourkela–769008, Odisha, India, Mobile: +91-7749004795, E-mail: nainsaxena@gmail.com

Shivanand Shankarrao Shirkole
PhD Research Scholar, Department of Food Process Engineering, National Institute of Technology at Rourkela, Rourkela–769008, Odisha, India, Mobile: +91-7064641005, E-mail: shivanandshirkole@gmail.com

Harshita Singh
PhD Research Scholar, Department of Biotechnology, Indian Institute of Technology Roorkee, Roorkee–247667, Uttarakhand, India, Mobile: +91-9829670161, E-mail: harshitasingh376@gmail.com

Harshvardhan Gowda Venkatachala
Professor, Department of Horticulture, University of Agriculture-Dharwad, Dharwad–580005, Karnataka, India, Mobile: +91-9535069142, E-mail: hvgv.117@gmail.com

Bharti Verma
PhD Research Scholar, Department of Chemical Engineering, Indian Institute of Technology-Roorkee, Roorkee–247667, Uttarakhand, India, E-mail: bverma@ch.iitr.ac.in

ABBREVIATIONS

µg/mL	microgram per milliliter
AFM	atomic force microscopy
Ag	silver
AgCl	silver chloride
AgNPs	silver nanoparticles
AL	alkaline lignin
Au	gold
Au-MSN	gold mesoporous silica nanoparticles
BAW	bulk acoustic wave
BHA	butylated hydroxyl anisole
BHT	butylated hydroxy toluene
BP	brake power
BSA	bovine serum albumin
BSFC	brake specific fuel consumption
BTE	brake thermal efficiency
Ca	calcium
CAC	Codex Alimentarius Commission
CDH	cellobiose dehydrogenase
CeNTAB	center for nanotechnology and advanced biomaterials
CFB	corrugated fiber board
CHx	cholesterol oxidase
Cl	chlorine
CLA	conjugated linoleic acid
CNP	carbon-based nanoparticles
CNS	central nervous system
CNT	carbon nanotube
CO	carbon mono oxide
CO_2	carbon dioxide
CP	conducting polymer
CPP	casein phospho-peptide
CQDs	carbon quantum dots
Cr(VI)	hexavalent chromium

Cu	copper
CV	coefficient of variation
CVD	chemical vapor deposition
CW	cellulose whiskers
DHA	docosahexaenoic acid
DHS	dynamic headspace
DLS	dynamic light scattering
DMSO	di-methyl-sulfoxide
DNA	deoxyribonucleic acid
DO	dissolved oxygen
DP	degree of polymerization
EFSA	European Food Safety Authority
ENCC	electro-sterically stabilized nanocrystalline cellulose
ENPs	enzyme nanoparticles
EPA	eicosapentaenoic acid
EPA	Environmental Protection Agency
EPR	enhanced permeability and retention
EU	European Union
EVOH	ethylene-vinyl alcohol
FAE	fatty acid esters
FAO	Food and Agriculture Organization
FDA	Food and Drug Administration
Fe	iron
Fe_3O_4	iron oxide
FSANZ	Food Standards Australia and New Zealand
FSMS	food safety management system
FTIR	Fourier transform infrared
g/L	gram per liter
GAP	good agricultural practices
GMP	good manufacturing practices
GOI	Government of India
GOx	glucose oxidase
HA	hyaluronan
HACCP	hazard analysis critical control point
HAP	hydroxylapatite nanorods
HC	hydrocarbon
HCA	heterocyclic amines
HCA	hierarchical cluster analysis

HCNC	hairy cellulose nano-crystalloid
HDPE	high-density polyethylene
HPH	high pressure homogenization
HPLC	high-performance liquid chromatography
HRP	horseradish peroxidase
ISI	Indian Standard Institution
ISO	International Organization for Standardization
JRC	Joint Research Center
$KMnO_4$	potassium permanganate
L-b-L	layer-by-layer
LCB	lignocellulose biomass
LDA	linear discriminant analysis
LDPE	low-density polyethylene
LHRH	luteinizing hormone-releasing hormone
LNPs	lignin nanoparticles
LPMO	lytic polysaccharide monooxygenase
M	molarity
MAP	modified atmospheric packaging
mg/mL	milligram per milliliter
MMT	montmorillonite
MNPs	magnetic nanoparticles
MOS	metal-oxides sensors
MOSFET	metal oxide semiconductor field effect transistors
mRNA	messenger ribonucleic acid
MW	multiple-walled
MWCNT	multi walled carbon nanotube
$MWCNT-ZrO_2$	multiwall carbon nanotube-zirconia nano hybrid
NCCP	national codex contact point
NF	nano filtration
NFC	nano fibrillated cellulose
NLCs	nanoliposomes
nm	nanometer
NO	nitric oxide
NO_2	nitrogen dioxide
NPC	nano-powdered chitosan
NPES	nano-powdered eggshell supplemented
NPPS	nano-powdered pea nut sprout
NPs	nanoparticles

NPS	nano-powdered peanut sprout
O_2	oxygen
P&T	purge and trap
PAH	polycyclic aromatic hydrocarbons
PCA	principal component analyses
Pd	palladium
PET	polyethylene terephthalate
PLA	polylactic acid
PM	particulate matter
PP	polypropylene
ppm	parts per million
PPP	public-private-partnership
PS	polystyrene
Pt	platinum
PTFE	polytetrafluoroethylene
PVC	polyvinyl chloride
R&D	research and development
RFID	radio frequency identification
RO	reverse osmosis
ROS	reactive oxygen species
Rsv	resveratrol
Se	selenium
SEM	scanning electron microscopy
SHS	static headspace
SiO_2	silicon dioxide
siRNA	small interfering RNA
SLN	solid liquid nanoparticles
SNCC	stabilized nanocrystalline cellulose
SnO_2	stannic oxide
SO_4	sulfate
SSOP	sanitation standard operating procedures
SW	single-walled
SWCNT	single walled carbon nanotube
TDS	thiamine di-lauryl sulfide
TEM	transmission electron microscopy
TiO_2	titanium dioxide
UF	ultra-filtration
UHT	ultra-heat treatment

US FDA	United States Food and Drug Administration
US HHS	United States Department of Health and Human Services
USDA	United States Department of Agriculture
USEPA	United States Environmental Protection Agency
UV	ultraviolet
V/s	volt per second
w/o	water into oil emulsion
WHO	World Health Organization
WO_3	tungsten trioxide
Zn	zinc
ZnO NPs	zinc oxide nanoparticles
ZnO	zinc oxide
ZnS	zinc sulfide

PREFACE

Nanotechnology is a small wonder. New discoveries at nano-scale have projected the possibility of new materials with improved functionality and architecture. In last decade, nanotechnology has rooted from soil to food in rapid phase. Application of nanotechnology in agriculture and its allied fields has improved the quality of respective field's outcome. Interaction between academics and enterprises is critical to understanding the market and laboratory constraints. Expectations of nanoscience have not yet been fulfilled, resulting in bioaccumulation and health and environmental concerns, which are yet to be explored.

This book mainly covers current scientific evidence on the basic concepts and terms associated with nanoscience technology in food safety and quality. The findings reported in this book can be useful for a long-term scientific vision of nanotechnology to enable and ensure its potential utilization in the food sector for food safety and health policy decisions.

Upoln searching the literature, one can find many specialized articles and books on nanotechnology and its application in different sectors. While they can be useful, a frequent limitation is that they tend to be imprecise and also often fail to identify and elaborate what are most likely to occur in food products. This book volume attempts to illustrate various aspects of nanotechnological application for different sectors.

This book has several potential users. It can be a reference book for those students who are taking for the first time college- or university-level courses on agriculture engineering, nanotechnological science, dairy science, veterinary science, pharmaceutical science, food safety, or quality assurance. This book will be further useful to engineering, pharma and food industry quality officers or employees who want to excel in their routine work.

This book is organized in such a way that each chapter treats one major application of nanotechnology for the agricultural processing, health, food safety, and bioprocess industries through various means. Therefore, this book is designed to provide more insights into utilization of nanotechnology in agriculture, dairy science, and food science.

In first part of book, the authors have introduced nanoscience application in the agriculture and horticulture sectors. The second part is dedicated to food science, where application of nano-formulations for food safety and for iron fortification of dairy food matrix, are discussed. In third part of the book, applications of nanotechnological approaches for bioprocess engineering were discussed in detail, which includes biofuel generation, wastewater treatment and enzymatic nanoparticles applications, and fabrication processes.

As a comprehensive compilation of recent developments of nano-technology in agriculture, dairy, and food science, this book provides an understanding of nanotechnology concepts and the critical issues in their respective areas.

This book comes under book series *Innovations in Agricultural & Biological Engineering.* This book volume is a treasure house of information and an excellent resource for those working in the areas of wastewater treatment, nano-formulation for target drug delivery, bioprocessing, agricultural quality control, and safety of food products during production, processing, and transportation in any food industry and will boost their confidence in the area of safety and quality aspects of food products.

We are grateful to all the authors for their expertise, commitment, and dedication.

We would also like to thank our editorial and production staff at Apple Academic Press, Inc. for their valuable help and advice throughout this project.

We also request our readers provide their fruitful suggestions for this volume, which will be helpful to improve the next edition.

Also, we would like to thank our families, for their continuous support throughout.

It is a lesson that motivates us in everything we do.

—Editors

PART I
Role of Nanotechnology in Agricultural Engineering

CHAPTER 1

ROLE OF NANOTECHNOLOGY: EMERGING PATH FROM SOIL TO FORK

LOHITH KUMAR DASARAHALLI-HULIGOWDA,
GAJANAN GUNDEWADI, and VIJAY S. RAKESH REDDY

ABSTRACT

The nanoemulsions, nanoparticles (NPs), nanocomposites, and nanosized powders provide a wide range of applications in food and agriculture processing. Nanocomposites, nanoemulsions, and nanocomposites are being used in packaging materials to enhance the packaging matrix barrier properties and also as antimicrobials. This chapter reviews different aspects of nanotechnology in agriculture and food sector.

1.1 INTRODUCTION

Nanoscale technology, engineering, and science can be described as sets of fundamental information and facilitating technologies as a consequent of efforts to recognize and regulate the functions and properties of matter at the nanoscale dimension. Nanotechnology is providing sustainable processing techniques in food and agriculture sector. It has proved its existence with numerous potential applications such as nanosensors, nano-pesticides, nano-fertilizers, nanocomposites, and nanoencapsulation [14, 16, 18, 37].

Dimensions of nanoparticles (NPs) (length and aspect ratio) show a discrepancy based on its end application. Nanofood is an edible matrix that uses nanotechnology tools, techniques or engineered nanomaterials that have been added during production (cultivation) and processing (post-harvest management, packaging, mixing, and homogenization) to enhance the quality and safety. Generally, NPs are characterized based on their characteristic dimension, i.e., less than 100 nm. The key innovative technology of nanotechnology holds in fabrication of nanosensors to detect food analytes such as pathogens, gases, organic compounds) [51, 63]. The production of NPs includes large scale chemical bonds and materials through bottom-up/top-bottom technology (Figure 1.1).

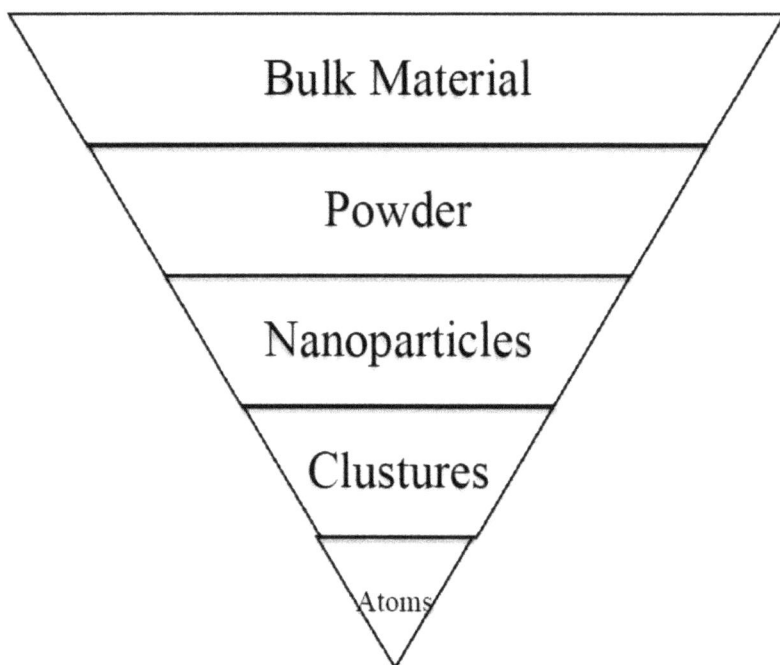

FIGURE 1.1   Illustration of bottom-up/top-down technology for preparation of nanoparticle.

Physicochemical (mass ratio, surface area), optical, and magnetic properties of such nanomaterials significantly different from their macroscale counterparts. The efficacy of the smaller NPs synthesized in the range of 10–12 nm is unstable; hence these NPs can be coated with polymer materials to retain its stability. However, reduction in the size of the particle enhances the bioactive efficacy when compared to the same particle's micro or macro size [42, 45].

Size of NPs generally relates to its functionality; reduction in size increases surface area, is desirable for purposes such as flavor release, increased water absorption, increased rate of catalysis and bioavailability. For instance, decrease in the particle size of *Chenopodium formosanum* (grain used as natural source of food coloring in Taiwan and rich in betanin) to 20–75 nm using nanogrinding with stabilized zirconia bead increased its radical scavenging activity and phenolic content compared to its granular form and microparticles. Nevertheless, the pigments in NPs are not stable compared to its microparticles matrix [72].

Both natural and synthetic nanomaterials are used in agriculture and food sector. Engineered nanocomponents are used as raw materials for the fabrication of nanoscale particles. The NPs are synthesized by organic, inorganic, combined materials and by surface functionalized materials vary with the mode of applications. For instance, carbon-based materials such as fullerenes, clay, carbon nanotubes (CNTs) and carbon black can be used in the synthesis of inorganic metal oxides, metals, and salts (Figure 1.2).

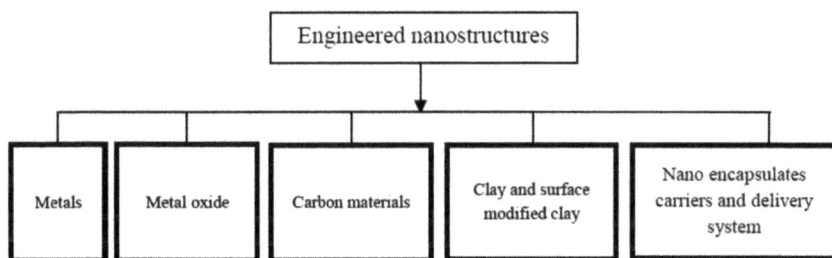

FIGURE 1.2   Illustration of different forms of nanostructures in agro-food industry.

The opportunities for utilization of NPs and other nanotechnology principles in agro-food industry are prodigious [29, 41, 66]. Biodegradable nanocomposite-based packaging matrices are gaining importance due to environmental concerns. Nanosized powders are being used to deliver and protect essential nutrients in food [69, 70]. Application of NPs in agriculture for plant protection and fast plant germination and production demonstrated potential future application of nanotechnology.

In addition, nano-sensors are being used to detect the pesticides residues in the produce and metal oxides and metal NPs have been utilized as pesticides to protect the crops. There are key concerns that need to be addressed before the communalization of nanotechnology in food and agriculture sector. The risk factor in terms of consumer acceptance in agriculture and food are high. Due to policy missteps and public concerns on genetically engineered food crops, the institution of nanotechnology in agriculture and food sector is the great challenge for both government and commercializing industry. In addition, few segments in public preferring organic food over nano-engineered foods and they are willing to pay premium price for food developed thorough organic approach and does not want their food to be 'bio-engineered or 'nano-engineered' according to Evan Micheleson and David Rejeski in their project of "Emerging Nanotechnologies [49]. Hence, at this point, there is a need of better understanding of nanotechnology future impacts and associated risk factors in both agriculture and food sector. This chapter focuses on applications and role of nanotechnology in agriculture and food processing.

1.2   NANO-REVOLUTION IN FOOD AND AGRICULTURE

Since the discovery of nanomaterials, several products with potential applications in the agri-food sector have been developed. These include nano-insecticides, and nano- emulsions for growth regulation, packaging materials, and pathogen detection devices based on antibodies, etc. Nanotechnology involves the application of agricultural inputs (fertilizers, insecticides, growth regulators, etc.), in nanometer-sized application or delivery systems in order to enhance the efficiency of application and utilization by the plant target, and to achieve more sustainable practices in agriculture and food areas. At present, agriculture is a highly chemical-intensive practice, primarily caused by the inefficiencies in the utilization and loss of fertilizers into water. By modifying the pattern of delivery and efficacy of agrochemicals through nanotechnology, plant protection, plant growth modification, enhanced stress tolerance, and environmental sustainability of agricultural production practices can be achieved.

Postproduction loss of horticultural products can be as high as 50% in developing countries because of inadequate and inefficient storage strategies. Nanotechnology has the potential to enhance the shelf-life, safety, and security of food through appropriate packaging technologies. Appropriate evaluations of safety and efficacy are required before food policies for nanotechnology and its applications in agriculture and food can be established [11, 15].

Nanostructures with unique characteristics are currently being used to enhance agriculture and food production. Research on engineered nanostructures is certainly illuminating novel applications that were once believed to be imaginary. Specifically, this cutting-edge technology applications result in higher productivity with nano-pesticides, nano-fertilizers, detection of pathogens, flavor, and other contaminants in food and increased product shelf life [58, 66]. However, some engineered nanostructures are found to be toxic to humans and environment.

Food packaging is the core area, where nanotechnology was initiated in food processing industry. Nanocomposites can be developed using multiple nanolayers of biopolymers or NPs to give the better strength and enhanced barrier properties [17, 27, 37, 52]. Packaging application of NPs is growing remarkably. It was observed that the zigzag pattern formed by nanostructures during film formation helps in prevention of gas diffusion (Figure 1.3). In addition, delivery of essential nutrients using nanotechnology principles is gaining importance.

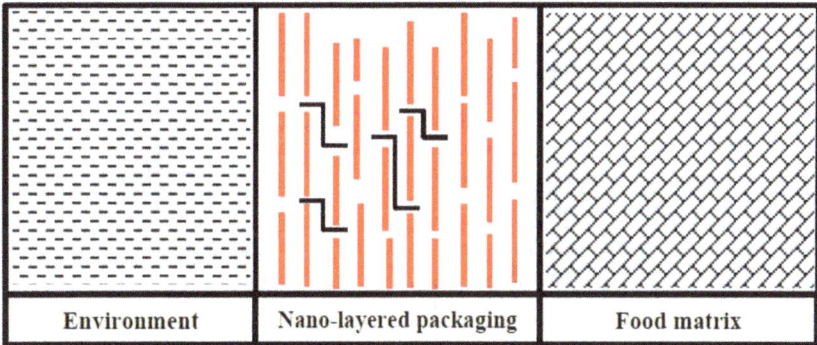

| Environment | Nano-layered packaging | Food matrix |

FIGURE 1.3 Illustration of gas-barrier system in multi-nanolayered packaging.

Nanoencapsulation methods such as spray drying, emulsification, extrusion, layer-by-layer (L-b-L) encapsulation and coacervation techniques are being used in food industry [30]. Nevertheless, nanostructures can be fabricated to block certain compounds in food matrix such as food allergens or harmful cholesterol from reaching certain parts of body. Currently, many leading food companies (such as: Unilever, Hershey, Kraft, Nestlé, and H. J. Heinz) are investing on nanotechnology-based research to overcome traditional processing constraints.

In addition to other fundamental applications of nanotechnology in agriculture, it is aimed to provide eco-friendly pesticides and fertilizers to increase the global food production. Nevertheless, engineered nanostructures can offer safe and efficient administration of pesticides, fertilizers, and herbicides and their controlled or targeted or stimuli-based release in plants. For instance, a pesticide formulated using tailored nanostructure can release or exhibit its properties only during targeted insect interaction [63]. In addition, the growth hormones dosage can be controlled using nanostructures which are used in enhancement of livestock production. However, nanosensors are being developed to detect and neutralize the animal pathogens in food chain [27].

1.3   NANOTECHNOLOGY IN AGRICULTURE

Agriculture is having more demand for improving efficiency in production of food. Application of nanotechnology in agriculture is comparatively underdeveloped. But nanotechnology has the potential to provide the solutions to these basic problems providing opportunities to use nano-fertilizers, nanoadditives, and pesticides that influence the crop health and yield [65].

At nanoscale level, the fertilizers will have increased surface area and lead to improved reactivity, faster dissolution, and uptake. Nano-additives include nutrients (Zn, pectin, rare earth oxides, Se, Fe, Ammonium Salts), pesticides, water retention materials, which are added to bulk products. Nano-fertilizers will be coated with polymers which help in controlled release of nutrients for examples nano clay and zeolites. Nevertheless, it is crucial to evaluate the benefits and risks of nanotechnology in agriculture nutrient management because agriculture is the main part in food chain and may influence the nanomaterial bioaccumulation in food web and chain [25].

Nanotechnology is the future innovative technology in agriculture sector which can deliver viable tools to traditional agriculture practice such as nano-pesticides and nanofertilizer. Advanced nano-forms of traditional agriculture fertilizer and pesticides provide targeted, controlled release and site-specific action. This can reduce the dosage required and reduces the residual contamination and eutrophication.

Utilization of nanoencapsulated metal as fertilizers and pesticides has been demonstrated their promising application in agriculture. Calcium, hydroxyapatite, iron, zinc oxide (ZnO), titanium dioxide gold NPs and

CNTs are examples of alternatives for traditional agro-chemicals. However, the potential toxic effects of gold, titanium dioxide, ZnO, iron oxide NPs consist of decreased growth, survival, fertility, and increased mortality in soil microorganisms. Hence, advanced, and more research is prerequisite in environment impact assessment of nanotechnology-based agro-chemicals before commercialization.

1.3.1 NANO-FERTILIZERS

Nanomaterials are also being used to develop environmentally friendly fertilizers to lower greenhouse gas emissions during crop production. Widely used carbon-based fertilizers like ammonium bicarbonate and urea in different soil types despite their efficiency are susceptible to decomposition and hydrolysis leading to elevated levels of byproducts such as nitrogen, ammonium carbonate and ammonia in soils. Further decomposition results in the accumulation of nitrate levels in soils leading to toxic effects and environmental damage. One such invention aims to reduce greenhouse gas emission whilst increasing fertilizer efficiency by reducing the decomposition of carbon containing fertilizers by incorporating carbon-nano constituents, nano-graphite, and carbon-colossal [11].

The nanofertilizer composition reduces gas emission by increasing the utilization of ammonium bicarbonate and thereby reducing the amount of fertilizer usage and increasing its efficiency. The NPs occupy cavities in the loosely assembled structure of ammonium bicarbonate which results in a more compact structure of increased stability in terms of its thermal and water properties. The extra rigidity provided by the inclusion of carbon-nano molecules decreases the susceptibility of ammonium bicarbonate to evaporation and increased usage of the material in the soil. Recently, Janmohammadi et al. demonstrated that application of nanofertilizer consists of nitrogen, potassium, phosphorous, iron, zinc, calcium, magnesium, manganese, molybdenum, and copper NPs enhance yield of potato [34]. Few examples for nano-fertilizers are provided in Table 1.1.

1.3.2 NANO-PESTICIDES

The utilization of nanotechnology practices in nano-pesticides development is comparatively novel and in the initial stages of development.

TABLE 1.1 Impact of Nano-Fertilizers on Agriculture Crops

Nano Fertilizer	Concentration	Crop Name	Application Method	Remarks	References
Chitosan loaded nitrogen, phosphorous, and potassium	500, 60, and 400 ppm	Wheat	Foliar	The results demonstrated that nanoparticles were taken up and transported through phloem tissues. Treatment of wheat plants grown on sandy soil with nano chitosan-NPK fertilizer induced significant increases in harvest index, crop index and mobilization index.	[1]
Gold nanoparticles	5 mg/L in 3.3 m² area	Red ginseng	Foliar	Use of a colloidal gold nanoparticle fertilizer improved the synthesis of ginsenosides in ginseng and enhanced the anti-inflammatory effects of red ginseng.	[36]
Nano-chelated micronutrients (nano-Zn, nano-Mn, nano-Fe)	–	Chickpea	Supplemental irrigation	Nano-chelated zinc and sulfur application increase seed yield, primarily due to an increase in the number of seeds per plant, secondary due to an increase in the harvest index, tertiary due to an increase in the number of pods per plant. High levels of sulfur fertilizer and nano-zinc can cause a significant increase in seed yield.	[33]
TiO_2, ZnO, Fe_2O_3	200 ppm	Barley	Foliar	Impact of nano zinc fertilizer was more prominent than iron. Foliar application of $nTiO_2$ positively affected some morpho-physiological characteristics like as days to anthesis, chlorophyll content and straw yield	[32]

Nanotechnology targets to decrease the bulk or undiscerning use of conventional pesticides in agriculture and their safety. The emphasis of ongoing research investigations was on the creation of nanoencapsulated pesticides with enhanced permeability, solubility, stability, and controlled release properties. These characteristics are primarily accomplished through either protecting the encapsulated functional compounds from early degradation (burst release) or enhancing their pest control efficacy for a longer period. In addition, the dosage of nanoencapsulated pesticides comparatively less and also it reduces the human exposure. However, lack of scientific evidences on the mechanism of interaction between nanoencapsulated pesticide and plants slowed down their potential application in nanopesticide delivery [11].

1.3.3 NANOPARTICLES (NPS) VERSUS SEED GERMINATION

Utilization of NPs on seed germination and in different stages of plant growth found to have both positive and negative impacts. Table 1.2 represents impact of NPs treatment on seed germination.

1.3.4 PLANT GROWTH

There are many investigations reported on nanomaterial-induced improvement in agronomic traits including yield, biomass content, and content of secondary metabolites by direct treatment in soybean, bitter melon, and rice indicating the ability of the nanomaterials in modifying genetic constitution of plants [39, 57, 64]. Nanomaterials have exhibited promise in targeted gene delivery for developing atomically modified plants a safer and acceptable strategy in contrast to genetic engineering. Interestingly, generational transmission of nanomaterials has been documented in rice and bitter melon [64]. The usage of these nanomaterials can ultimately land in our food cycle and so a careful study and analysis is pertinent regarding their usage before putting these materials in actual use.

The spurt in the research in this interdisciplinary field that involves primarily the fusion of nanotechnology and plant science may lead to the creation of a new field as Plantnanomics. Plant growth is characterized by increase in biomass of the germinated seeds. The length of roots and

TABLE 1.2 Impact of Nanoparticles Seed Treatment on Germination Potential

Nanoparticle	Seed	Concentration of Nanoparticles	Remarks	References
Carbon nanotubes	Tomato	10–40 µg/mL	Results demonstrated for the first time that carbon nanotubes can penetrate thick seed coat and support waster uptake inside seeds. The activated process of water uptake could be responsible for the significantly faster germination rates and higher biomass production.	[38]
	Barley, soybean, corn	50, 100, 250 µg/mL	Results demonstrated that carbon nanotubes can activate the expression of seed-located water channel genes (aquaporins) that belong to different gene families of aquaporins.	[43]
Gold nanoparticles (Au)	*Pennisetum glaucum*	20, 50 µg/mL	The uptake and penetration of dispersed nanoparticles through the seed coat might have created "nano-holes" resulted in improved germination. The use of lesser concentrations, slow, and minimal release of Au ions could be one of the reasons to have no major effect on germination of *P. glaucum* seeds. The highest percentage of seed germination was 86.66% and increased seedling length was observed at 50 µg/ml of gold nanoparticles followed by 20 µg/mL. The seed germination and seedling growth in this study was positively altered by nanoparticles.	[54]
Fe_3O_4	Cucumber	0–5,000 µg/mL	Germination was in increasing order from the concentrations of 100 µg/mL to 250 µg/mL followed by the decrease in germination with increase in concentration. This demonstrates that seeds experienced the stress at higher concentration nanoparticles.	[50]

TABLE 1.2 *(Continued)*

Nanoparticle	Seed	Concentration of Nanoparticles	Remarks	References
TiO$_2$	Cucumber	0–5,000 µg/mL	Seed germination was more inhibited by titanium dioxide than by iron oxide nanoparticles.	[50]
	Wheat	1–500 ppm	Treatment of TiO$_2$ at 1–10 ppm demonstrated accelerated germination and vigor in wheat seeds.	[23]
SiO$_2$	Tomato	8 g/L	Application of silicone di-oxide significantly enhanced germination potential of seeds. It is effective in yield and growth of crops. In addition, it shows the potential to use as fertilizers for the crop improvement.	[67]
Zero-valent iron nanoparticles	Ryegrass, barley, Flaxseed	0–5,000 mg/L	Zero-valent iron nanoparticles at low concentrations can be used without detrimental effects on plants and thus be suitable for combined remediation where plants are involved.	[19]
Silver nanoparticles (Ag)	Ryegrass, barley, flaxseed	0–100 mg/L	Silver nanoparticles inhibited seed germination at lower concentration, but showed no clear size dependent effects and never completely impeded germination. Thus, seed germination tests seem less suited for estimation of environmental impact of silver nanoparticles.	[19]
Hydroxylapatite nanorods (HAP)	Chickpea	0.5, 1.0, 1.5 mg/mL	The best performance was observed in presence of 1 mg/mL HAP nanorods. It enhanced seed germination and plant growth.	[7]

shoot, number of laterals, number, and size of leaves, total biomass, and yield represent the major growth parameters. Many studies on NPs-plant interactions have focused on these factors and reports of both enhanced and retarded growth have been documented. Most of the NPs applied to seedlings or plants as such are through the roots, which causes a kind of bias in determining their effects. The main reason behind this speculative assessment is that the movement of NPs through the plant tissues (translocation) has not been clearly understood, although there are a few reports available. Overall, the observed effects are linked to the nanoparticle interaction with roots, either promoting or blocking nutrient supply and subsequent translocation to higher tissues [64].

1.4   NANOTECHNOLOGY IN FOOD PROCESSING

Nanotechnology offers promising novel applications in almost all disciplines. Immense potential of nanotechnology has also recently touched the food industry. The doctrines of nanotechnology cleared the path to an unexplored science for studying individual NPs and their application in the food industry (Figure 1.4). Nano-foods provide a host of advantages such as improved shelf-life, enhanced bioavailability of health promoting bioactive compounds, and enhanced safety of food against spoilage factors. Current research in the area of food nanotechnology is mainly focused on designing of natural nanostructures for protecting nutrients in functional foods to improve their bioactivity and bio-accessibility at different environmental stresses. Naturally foods contain nanocomponents like proteins, carbohydrates, and lipids, which determine their properties.

Food processing conditions such as pH, temperature, pressure, ionic strength changes affect the naturally occurring nanostructure in food and result in structural changes at nano and micro level. Food spoilage during processing, transport, and packaging can be reduced to a greater extent by applying the concepts of nanotechnology. Nutraceuticals are more beneficial and possess no side effects, as mostly they are phytochemical compounds of natural resources. However, nutraceuticals retain less stability, efficacy, and bioavailability when they enter the human body system. Hence, to overcome such problems, nanotechnology was applied in the field of nutraceuticals.

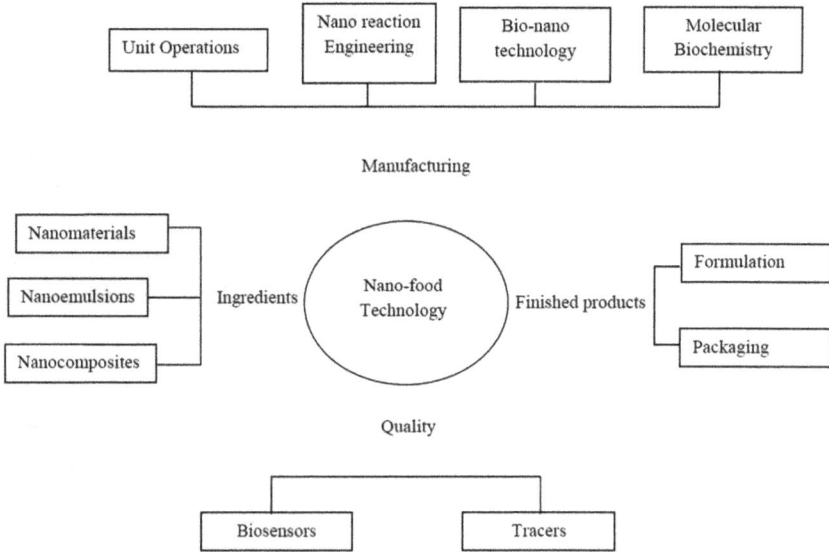

FIGURE 1.4   Illustration of application of nanotechnology in food processing.

1.4.1   FOOD COATING AND PACKAGING

Nanostructures are being used to create or recreate coatings and films with unique characteristics. Functional packaging is the new trend which includes encapsulation of functional molecules (antioxidants, antimicrobials) in film matrix. Emulsions, hydrocolloids, and biopolymers are being used to formulate food coatings and films. Few examples on biopolymer based multilayered films are provided in Table 1.3 on utilization of biopolymers to enhance the properties of food-grade films. Nanocomposites with biopolymer matrix enhance the filler-matrix interactions and its physical strength. Nano-clay is the most used composite in food packaging. For instance, polyvinyl alcohol, nylon, and polylactide are nano-clay based composites are being used with biopolymer matrices. Providing antimicrobial and antioxidant properties to food packaging materials was enabled through nanotechnology [21].

Nano-clay materials are often used to improve the gas barrier properties for oxygen or carbon dioxide (CO_2) while adding some mechanical strength, stiffness, heat resistance, and UV protection. These properties result from the platelet structure of nano-clays (e.g., montmorillonite (MMT) composed of plate-like particles of approximately 9 nm thick and 1 μm wide). Some packaging applications require a reduction in the number of bacteria that accumulates on the surface of the plastic package. A very common solution to this is to incorporate nano-silver in the plastic before molding or forming. Nano-silver acts as an antibacterial agent even when it is embedded inside a neutral matrix. Some ions leach out and form a very diluted solution of Ag+ ions on the surface of the packaging container; the Ag+ ions are responsible for the antibacterial effect.

TABLE 1.3 Examples of Food-Grade Biopolymers and Their Functional Properties in Film Formulation

Biopolymers	Remarks	References
Bacterial cellulose, zein fibers	Multilayer bacterial cellulose-zein films were formulated by incorporating electro spun zein fibers as the interlayers between bacterial cellulose films. Fabricated multilayer demonstrated enhanced resistance to water permeability.	[73]
Chitosan, nanoclay	Results demonstrated that inclusion of nanoclay in chitosan polymer-based films enhances the gas barrier and mechanical properties. This might be attributed to result of filling void spaces in chitosan matrix by nanoclay particles.	[53]
Polyhydroxybutyrate-co-valerate, zein, sodium alginate	Zein protein with cinnamaldehyde amalgamation was electro spun as interlayer between polyhydroxybutyrate-co-valerate layers.	[8]
Wheat gluten-hey protein isolate, wheat gluten-zein, wheat gluten-soy protein isolate	α-tocopherol was encapsulated in different matrices such as whey protein isolate, soy protein isolates and zein (produced by means of electro-hydrodynamic technique), which are directly applied as coatings on wheat gluten film. Bilayer matrix enhanced the barrier properties and protected the antioxidant activity of functional molecule during sterilization.	[20]
Whey protein, polyethylene terephthalate, polyethylene	Whey protein layer was added at intern of polyethylene and polyethylene terephthalate films. Multilayered films demonstrated enhanced barrier properties.	[13]

Generally, essential oils nanoemulsion, silver nanoparticles (AgNPs) are being used in food industry for the fabrication of antimicrobial film matrix. Recently, nano-fibers, composites, and NPs are used to create biodegradable packaging materials [24]. A nanolaminate consists of double or several layers of nanomaterial which are physically or chemically bound to each other (Figure 1.5). L-b-L deposition approach is one among the suitable methods for coating the interfacial films of charged surfaces which consists of multiple nanolayers made of various nanomaterials.

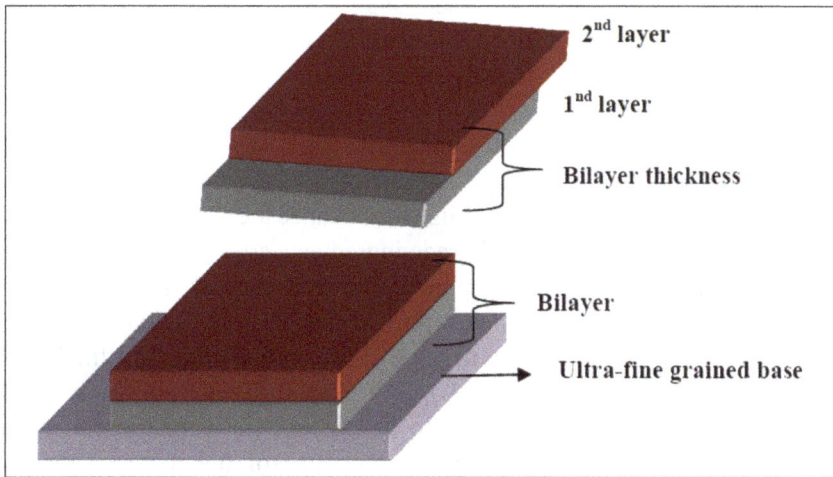

FIGURE 1.5   Illustration of nanolaminate used in food packaging.

The fundamental principle behind the L-b-L technology can be briefed as follows; when two individual opposite charged biopolymers are used to create the laminate, they attract each other during deposition over other polymer and enhancing the binding capacity between the layers. L-b-L technique helps in control of thickness, functional properties, charges, and barrier properties of laminates. For instance, charged biopolymers such as chitosan (cationic), pectin (anionic), proteins (possesses cationic behavior at a pH below isoelectric charge and anionic behavior at a pH above its isoelectric point) [56].

In addition to bio-polyelectrolytes (polysaccharides, proteins), colloidal particles (droplets, vesicles, micelles), charged lipids (surfactants, phospholipids), nanoemulsions are being used to create multiple layered nano-laminates. Factors such as soaking medium, number of nanolayers, preparation condition and nanolayers sequence can determine the functionality of nanofilms. For instance, their permeability to minerals, gases, water, organic substances; their mechanical properties such as flexibility, rigidity, wetting nature, swelling nature, brittleness; their sensitivity to ionic strength, temperature, and pH change can determine functionality of films [5, 44, 61].

In addition, encapsulation methods such as emulsification or matrix amalgamation can be used to encapsulate lipophilic and hydrophilic functional molecules in the packaging matrix. For instance, lipophilic functional molecules such as essential oil (antimicrobial property) can be incorporated into the film using emulsification technique and with the association of colloids (liposomes or micelles) functional activity could be enhanced at lower concentration. As a result, it demonstrates the possible ways to incorporate functional compounds such as antioxidants, flavors, colors, anti-browning agents and antimicrobials into the film matrix [55]. These functional molecules are utilized to enhance the shelf-life, safety, and quality of food matrix. All food-grade ingredients are used in manufacture of nanolaminates in foods by dipping and washing actions. The ingredients which are mostly used are proteins, polysaccharides, and lipid [4].

There are number of ways to control the composition and thickness of nanolayers coating, such as: (i) governing difference in the kind of adhering matters for the solution to be dipped, (ii) vary the complete methods for dipping, (iii) differ the sequence of the material that is launched into the different soaking solutions, (iv) diverge the liquid mixture or surrounding parameters used, such as dielectric constant, ionic strength, pH, and temperature. Electrostatic force, hydrophobic interactions, hydrogen bonding and thermodynamic properties are driven forces for the absorption between layers.

The impact of substrate surface properties such as roughness and topology can influence the barrier function of nanolayered laminates. Consequently, this would necessitate the fabrication of subsequent biopolymer nanolayers on the food matrix [60]. Although nanotechnology holds many promises for the development of innovative packaging for fruits and vegetables, there are some concerns about their potential toxicity due to lack of toxicological data of nanomaterials. Moreover, consumer acceptance of new packaging systems may vary in different marketplaces and demographics. Thus, the judicial selection of materials that comply with food regulations, along with gaining insight about consumer perception of the innovative packaging involved, are essential to ensure commercial viability.

1.4.2   SCOPE OF NANO-ADDITIVES IN FOOD FORTIFICATION

Growing world population demands for nutritional and quality food. Nutrients availability differs from place to place and food matrix to matrix. Utilization of nanotechnology in the production and protection of artificial colors, flavors, preservatives to enhance the food shelf life could be the only way to attain sustainable food processing. For instance, synthetic nano-amorphous silica used as clarifying agent and anti-caking agent [9]. Another important coloring agent is titanium dioxide. It is used as a pigment to increase the turbid color of specific edible, such as milk products and candies. Titanium dioxide leads to the inactivation of food born pathogen, in its nano form or when combined with elements such as nickel oxide and cobalt. European Food Safety Authority (EFSA) has certified the safety of titanium dioxide and can be used as a coloring agent in food and feed [52].

Food additives such as iron NPs are recognized as health enhancing compounds, i.e., it is an essential mineral required for body. In a recent study, it was evidenced that, β-lactoferrin can be used to as vector protect the iron from oxidation [47]. The physio-chemical characteristics such as solubility and bioavailability of iron particles in which are weak acid-soluble can be boosted by particle size reduction in their primary stage and thereby expanding their specific surface area [31].

Nano-delivery systems are utilized in enhancing the efficacy of active food additives. To protect the bioactive compounds like functional carbohydrate (prebiotics), protein (peptides) and active lipids (ω-3 fatty acids) nano-encapsulation has been employed. It involves the incorporation, absorption, and also diffusion of functional NPs or compound in specific targeted area with response to stress factors such as pH change, enzyme action and temperature. Nanoencapsulation of food additives helps in overcome of stability and solubility problems (For instance, solubilizing a hydrophobic compound in hydrophilic matrix and vice versa). Examples for nanoencapsulation are provided in Table 1.4.

1.4.3   NANO-NUTRACEUTICALS

Foods are structurally composed of a complex assembly of nanosized elements with varying physical and chemical characteristics that

TABLE 1.4 Examples of Nanoencapsulation of Functional Compounds

Nano-Encapsulation Matrix	Functional Compound	Remarks	References
Nanoemulsion	Curcumin	The bioaccessibility of curcumin appeared to be slightly higher in conventional emulsions than in nanoemulsions, but nanoemulsions had much better physical stability.	[2]
Nanoparticles	Vitamin D$_3$	Encapsulation of hydrophobic nutrients in zein/CMCS complex nanoparticles would achieve the controlled release property and improve the stability of labile nutrients	[46]
Liposomes	Fish oil (Rich in cis-4,7,10,13,16, 19-docosahexaenoic acid (DHA) and cis-5,8,11,15,17-eicosapentaenoic acid (EPA).	Nano-liposome encapsulation resulted in a significant reduction in acidity, syneresis, and peroxide value while increasing DHA and EPA stability.	[26]
Solid lipid nanoparticles	Polyphenolic and anthocyanin compounds	Encapsulation of the hydrophilic bioactives at high entrapment efficiency (exceeding 90%) into nano-sized spherical particles increased the apparent short-term accelerated stability of anthocyanins and polyphenolic compounds against relatively high pH and temperatures.	[59]

determine the overall stability and properties of the interfacial colloidal mixture containing lipids, proteins, carbohydrates, artificial additives, and surfactants. Controlling how NPs, form assemble and interact with each other using nanotechnological tools provides a means to design foods from scratch. The potential of 'nanotechnology' in enhancing food products was recognized a decade ago by leading food industrialists, academia, and governmental bodies who combined their efforts as a consortium to explore the potential of introducing nano-based materials in foods [61, 75].

Incorporation of functional ingredients in functionally coated nanomaterials to drive the development of functional nanofoods is another area of major active interest in the food industry. In pursuit of this technology, a number of coating methods like physical and chemical vapor deposition (CVD), pyrolysis, sol-gel processes and supercritical CO_2 have been used. Suspension of NPs in the polymer or organic solvent remains largely insoluble and introduction of the supercritical fluid results in encapsulation or coating of the precipitated NPs resulting in modified surface properties defined by the polymer of interest [6, 10, 48]. The method has the added advantage of having greater control over the coating process with increased encapsulation efficiency and efficacy of NPs.

Nanotechnology is also playing a role in enhancing the delivery of probiotics. Delivery of beneficial microorganisms that increase nourishment through natural biological processes has also benefited from nanoencapsulation technology. Examples of favorable bacteria used in probiotics are *Lactobacillus salivarius, Lactobacillus acidophilus, Saccharomyces boulardii, Saccharomyces thermophilus* and *Bifidobacterium* species that aid the digestion of food, increase energy storage and fermentation of sugars, help prevent tumor formation, stimulate the production and release of vitamins and antibiotics, and inhibit the development of pathogenic conditions such as those associated with infections, inflammation, and heart disease [22].

They are only beneficial in humans if their useful functions and properties are not affected by processing methods and are able to reach specific locations in the body without being destroyed. To meet these requirements, the encapsulation of probiotics has received much attention in the food industry. An example of nano coated probiotic bacteria resistant to degradation by gastric acid as a food supplement has been developed [71]. Researchers have developed a protective coating for lactic acid bacteria which shows increased heat, acid, and bile resistance and overall stability. The triple coated bacteria comprise of protein coat (e.g., soya isolate) in the presence of a protease, polysaccharides (e.g., xanthan gum) and finally coated with NPs (e.g., solid lipid nanoparticles). The bacteria remain intact in bile, nutritionally active and can be added directly as food supplements in a variety of products such as fermented milk [12].

1.5   CHALLENGES FOR COMMERCIALIZATION OF NANOTECHNOLOGY

The crucial steps taken by government agencies and industries towards commercialization of nanotechnology are very less. As we discussed in earlier section, few segments in public demands organic foods and organic food sector are associated with a tagline 'GMO-FREE' and currently, it is the fastest growing area in food industry. Nevertheless, food industries do not want circumstances in which customers paying attention to food products for the reason that they claim to be 'NO-NANO' or 'NANO-FREE.' In addition, the progression of organic food industry specifies that industries may gain real economic benefit with affirmative food package labeling ('NO-NANO' or 'NANO-FREE') that goes further than just taking a situation against nano-engineered foods. In addition, nanofood or nanoparticle amalgamated food matrices biological fate is unknown in many cases. Based on these views, few challenges are briefed in subsections.

1.5.1   PUBLIC PERCEPTION AND INDUSTRIES VANGUARD

As other innovative technologies incorporation in food and agriculture processes, nanotechnology magnificently reached the top among many innovations. However, public perception greatly influences the future of any technology and its application. The dearth of nanotechnology information and its risk assessment in agri-food matrices may compli-cate the public perception. Both food and agribusiness are at the front-line of commercializing nano-based innovative technologies, whilst, their successes or failures could disturb future commercialization of nano-products.

1.5.2   ENVIRONMENTAL CONCERNS

The major risk associated between environment and nanotechnology is the initial phases of nanotechnology pertaining to their safety and unregulated product formulation and development. In particular, attention should be given to nanomaterials as they have unusual properties and their unique characteristics make them a double-edged sword. For instance, they can be engineered to yield enhanced properties and special benefits; however, it can also demonstrate unexpected environmental and toxicological effects.

1.5.3   REGULATORY SYSTEMS

Regulatory concerns are the major gap between the lab invention and commercialization. There are many food additives which are used in food formulations which may vary in the dosage. Nano-additives are made by metal and its oxides which are partially soluble or insoluble and indigestible in the body. Utilization of such additives may bioaccumulate in body imposing health concerns. An important challenge fronting the regulatory systems in the nanotechnology-based products are, a greater number of sectors giving attention to nanotechnology-based product development. To recognize the vulnerability and make sure that nanotechnology or nanoscience does not 'drop through the flaws' a double overview method should be implemented, one that necessities the research on long-term and short-term risk associated with the nanotechnology, while, other should look into the potential transformations and shifts in nanotechnology implication that may occur in the future [3].

1.5.4   TOXICOLOGY AND RISK ASSESSMENT

Nano forms of various organic, inorganic materials are being used in the formulation of new products. However, if their interactions with the biological systems are altered, it may become toxic and causes abnormal events in the system, Toxicology measurement at nanoscale is the new step towards sustainable nanotechnology, which give emphasis to utilization of nano-structures in food and agriculture sector. Although consumption of passive nanostructures is very less compared to active nanostructures, yet their toxicity level is also assessed to be low, however, these nanostructures may accumulate in host and it may explicit the adverse effect later. In addition, at one stage of accumulation, nanostructures may become an eco-hazard [14].

For instance, nanostructures such as quantum dots, CNTs, metal, and metal oxides have adverse effects when their concentration exceeds critical concentration which is even in low level. There is no perfect outlook on which parameter(s) should be measured as a most suitable measure of evaluating exposure (number/mass/surface area/size). The inverse correlation between surface area and particle size, it is imperative that, concentration, and its effect relationship can be established as a function of nanoparticle size (sometimes charge) and surface charge instead of mass units [62]. The factors that need to be considered in the risk assessment of new forms of NPs associated with a process or product are shown in Figure 1.6.

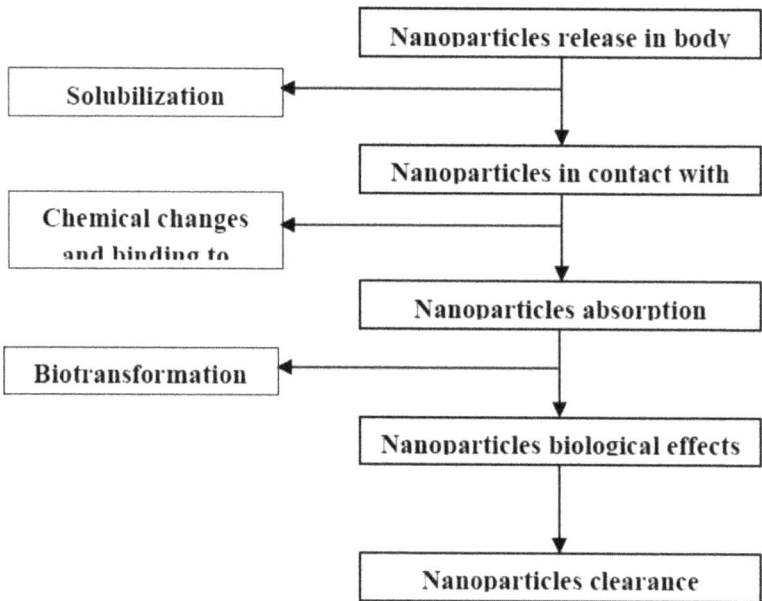

FIGURE 1.6   Toxico-kinetics of nanoparticles in body.

1.6   REGULATIONS FOR NANOTECHNOLOGY: AN INSIGHT INTO THE FUTURE

As the challenge in assessing the safety of nanofoods and nano-packaging becomes more complex with the arrival of novel nanomaterials for use in the food industry, greater cooperation is required to ensure that human and environmental concerns are not compromised as new products are released. Therefore, the pace of introducing food technology must be sufficiently slowed to allow potential risks to be identified and assessed for a safer future. This essentially means that innovation must be balanced by regulatory guidelines through the availability of reliable and robust risk-assessment tools which currently do not exist for nanofoods [28, 35, 68, 70].

With the introduction of NPs in food matrix, methods for the identification and quantification of nanosized particles in food matrices are being developed for future regulatory testing in terms of the distribution and migration of engineered particles in food stuffs. These devices will be necessary to instill public confidence in nanoparticle-based products to ensure that shelf products are quality assured and safe.

A lack of analytical assessment tools in this area has been met by efforts to establish methodologies and instrumentation for food analysis [35]. However, this area will require further attention as regulatory authorities may implement restrictions on the use of NPs as food components in the future. One such invention involves an apparatus designed to assess engineered NPs in a substrate. The analysis describes a non-invasive and non-destructive means of detecting, measuring the size distribution, identifying, and quantifying food embedded NPs.

The method relies on the use of a nanoparticle loaded substrate used as a reference material and combines imaging and analytical systems based on screening techniques such as dynamic light scattering (DLS) (particle size distribution), mass spectroscopy (test material composition), surface plasmon resonance (mass fraction), spectroscopy (nanoparticle sizing), positron emission tomography (radio tracing), optical emission spectrometry (trace level elemental analysis) for collecting regulatory information [40, 74].

1.7 SUMMARY

This chapter reflects on the utilization of nanotechnology in agriculture and food sector, such as: nano-fertilizers, nano-pesticides, seed germination, nanoencapsulation of functional molecules and nanocomposite formulation. In addition, few examples have been given to understand the functionalities of NPs and nanostructures in agriculture and food in

Sections 1.3 and 1.4. Formulation of nano-layered packaging matrices was discussed in Section 1.4. Section 1.5 presents challenges for nanotechnology implementation and its commercialization in both agriculture and food sector. Regulatory frameworks and factors to be considered during risk assessment of nanotechnology (charge, size, and surface area) are discussed in Section 1.6.

KEYWORDS

- amalgamation
- nanoencapsulation
- nano-fertilizer
- nano-nutraceuticals
- nanoparticle toxicity
- plantnanomics
- toxico-kinetics
- xanthan gum
- zeolites

REFERENCES

1. Abdel-Aziz, H. M., Hasaneen, M. N., & Omer, A. M., (2016). Nano chitosan-NPK fertilizer enhances the growth and productivity of wheat plants grown in sandy soil. *Spanish Journal of Agricultural Research, 14*(1), 8. Article ID: 0902.
2. Ahmed, K., Li, Y., McClements, D. J., & Xiao, H., (2012). Nanoemulsion and emulsion-based delivery systems for curcumin: Encapsulation and release properties. *Food Chemistry, 132*(2), 799–807.
3. Amenta, V., Aschberger, K., Arena, M., & Bouwmeester, H., (2015). Regulatory aspects of nanotechnology in the Agri/feed/food sector in EU and non-EU countries. *Regulatory Toxicology and Pharmacology, 73*(1), 463–476.
4. Appendini, P., & Hotchkiss, J. H., (2002). Review of antimicrobial food packaging. *Innovative Food Science & Emerging Technologies, 3*(2), 113–126.
5. Arora, A., & Padua, G., (2010). Review: Nanocomposites in food packaging. *Journal of Food Science, 75*(1), R43–R49.
6. Bagchi, D., Lau, F. C., & Ghosh, D. K., (2010). *Biotechnology in Functional Foods and Nutraceuticals* (p. 591). Boca Raton - FL: CRC Press.

7. Bala, N., Dey, A., Das, S., Basu, R., & Nandy, P., (2014). Effect of hydroxyapatite nanorod on chickpea (*Cicer arietinum*) plant growth and its possible use as nano-fertilizer. *Iran. J. Plant Physiol, 4*(3), 1061–1069.

8. Cerqueira, M. A., Fabra, M. J., & Castro-Mayorga, J. L., (2016). Use of electrospinning to develop antimicrobial biodegradable multilayer systems: Encapsulation of cinnamaldehyde and their physicochemical characterization. *Food and Bioprocess Technology, 9*(11), 1874–1884.

9. Chaudhry, Q., Scotter, M., & Blackburn, J., (2008). Applications and implications of nanotechnologies for the food sector. *Food Additives and Contaminants, 25*(3), 241–258.

10. Chen, L., Remondetto, G. E., & Subirade, M., (2006). Food protein-based materials as nutraceutical delivery systems. *Trends in Food Science & Technology, 17*(5), 272–283.

11. Chhipa, H., (2016). Nanofertilizers and nanopesticides for agriculture. *Environmental Chemistry Letters, 15*(1), 1–8.

12. Chhipa, H., & Joshi, P., (2016). Nanofertilizers, nanopesticides and nanosensors in agriculture. In: Ranjan, S., Dasgupta, N., & Lichtfouse, E., (eds.), *Nanoscience in Food and Agriculture - I* (Vol. 20, pp. 247–282). Sustainable agriculture reviews. Cham: Springer.

13. Cinelli, P., Schmid, M., Bugnicourt, E., Coltelli, M. B., & Lazzeri, A., (2016). Recyclability of PET/WPI/PE multilayer films by removal of whey protein isolate-based coatings with enzymatic detergents. *Materials, 9*(6), 473–480.

14. Cushen, M., Kerry, J., Morris, M., Cruz-Romero, M., & Cummins, E., (2012). Nano-technologies in the food industry-recent developments, risks and regulation. *Trends in Food Science & Technology, 24*(1), 30–46.

15. Dasarahalli-Huligowda, L. K., Goyal, M. R., & Suleria, H. A. R., (2019). *Nanotechnology Applications in Dairy Science: Packaging, Processing, and Preservation* (1st edn., p. 275). Oakville - ON: Apple Academic Press Inc.

16. Lohith, K. D. H., & Preetam, S., (2018). Potential of nanotechnology in dairy processing: A review. In: Megh, R. G., (ed.), *Sustainable Biological Systems for Agriculture* (pp. 55–79). Oakville - ON: Apple Academic Press Inc.

17. Duncan, T. V., (2011). Applications of nanotechnology in food packaging and food safety: Barrier materials, antimicrobials and sensors. *Journal of Colloid and Interface Science, 363*(1), 1–24.

18. Duran, N., & Marcato, P. D., (2013). Nanobiotechnology perspectives. role of nanotechnology in the food industry: A review. *International Journal of Food Science & Technology, 48*(6), 1127–1134.

19. El-Temsah, Y. S., & Joner, E. J., (2012). Impact of Fe and Ag nanoparticles on seed germination and differences in bioavailability during exposure in aqueous suspension and soil. *Environmental Toxicology, 27*(1), 42–49.

20. Fabra, M. J., López-Rubio, A., & Lagaron, J. M., (2016). Use of the electrohydrody-namic process to develop active/bioactive bilayer films for food packaging applications. *Food Hydrocolloids, 55*, 11–18.

21. Farhoodi, M., (2016). Nanocomposite materials for food packaging applications: Characterization and safety evaluation. *Food Engineering Reviews, 8*(1), 35–51.

22. Favaro-Trindade, C., Heinemann, R., & Pedroso, D., (2011). Developments in probiotic encapsulation. *CAB Rev., 6*, 1–8.

23. Feizi, H., Rezvani, M. P., Shahtahmassebi, N., & Fotovat, A., (2012). Impact of bulk and nanosized titanium dioxide (TiO_2) on wheat seed germination and seedling growth. *Biological Trace Element Research, 146*(1), 101–106.

24. Ferrer, A., Pal, L., & Hubbe, M., (2016). Nanocellulose in packaging: Advances in barrier layer technologies. *Industrial Crops and Products, 95*, 574–582.

25. Fraceto, L. F., Grillo, R., De Medeiros, G. A., Scognamiglio, V., Rea, G., & Bartolucci, C., (2016). Nanotechnology in agriculture: Which innovation potential does it have? *Frontiers in Environmental Science, 4*, 20–28.

26. Ghorbanzade, T., Jafari, S. M., Akhavan, S., & Hadavi, R., (2017). Nano-encapsulation of fish oil in nano-liposomes and its application in fortification of yogurt. *Food Chemistry, 216*, 146–152.

27. Gruère, G. P., (2012). Implications of nanotechnology growth in food and agriculture in OECD countries. *Food Policy, 37*(2), 191–198.

28. Hassellöv, M., Readman, J. W., Ranville, J. F., & Tiede, K., (2008). Nanoparticle analysis and characterization methodologies in environmental risk assessment of engineered nanoparticles. *Ecotoxicology, 17*(5), 344–361.

29. Hati, S., Makwana, M. R., & Mandal, S., (2019). Encapsulation of probiotics for enhancing the survival in gastrointestinal tract. In: Huligowda, L. K. D., Goyal, M. R., & Suleria, H. A. R., (eds.), *Nanotechnology Applications in Dairy Science: Packaging, Processing, Preservation* (pp. 225–241). Oakville - ON: Apple Academic Press Inc.

30. Huang, Q., Yu, H., & Ru, Q., (2010). Bioavailability and delivery of nutraceuticals using nanotechnology. *Journal of Food Science, 75*(1), R50–R57.

31. Jackson, L. S., & Lee, K., (1991). Microencapsulated iron for food fortification. *Journal of Food Science, 56*(4), 1047–1050.

32. Janmohammadi, M., Amanzadeh, T., Sabaghnia, N., & Dashti, S., (2016). Impact of foliar application of nano micronutrient fertilizers and titanium dioxide nanoparticles on the growth and yield components of barley under supplemental irrigation. *Acta Agriculturae Slovenica, 107*(2), 265–276.

33. Janmohammadi, M., & Sabaghnia, N., (2016). Statistical assessment of the impact of nano-chelated elements and sulfur on chickpea production under supplemental irrigation. *Agriculture & Forestry/Poljoprivreda i Sumarstvo, 62*(2), 263–274.

34. Janmohammadi, M., Sabaghnia, N., Nouraein, M., & Dashti, S., (2016). Responses of potato (*Solanum tuberosum* L. var. *agria*) to application of bio, bulk and nano-fertilizers. *Annales Universitatis Mariae Curie-Sklodowska, Sectio C–Biologia, 70*(2), 57–65.

35. Kandlikar, M., Ramachandran, G., Maynard, A., Murdock, B., & Toscano, W. A., (2007). Health risk assessment for nanoparticles: A case for using expert judgment. *Journal of Nanoparticle Research, 9*(1), 137–156.

36. Kang, H., Hwang, Y. G., Lee, T. G., Jin, C. R., Cho, C. H., Jeong, H. Y., & Kim, D. O., (2016). Use of gold nanoparticle fertilizer enhances ginsenoside contents and anti-inflammatory effects of red ginseng. *Journal of Microbiology and Biotechnology, 26*(10), 1668–1674.

37. Kaya-Celiker, H., & Mallikarjunan, K., (2012). Better nutrients and therapeutics delivery in food through nanotechnology. *Food Engineering Reviews, 4*(2), 114–123.

38. Khodakovskaya, M., Dervishi, E., Mahmood, M., Xu, Y., Li, Z., Watanabe, F., & Biris, A. S., (2009). Carbon nanotubes are able to penetrate plant seed coat and dramatically affect seed germination and plant growth. *ACS Nano, 3*(10), 3221–3227.
39. Kole, C., Kole, P., Randunu, K. M., Choudhary, P., Podila, R., Ke, P. C., Rao, A. M., & Marcus, R. K., (2013). Nanobiotechnology can boost crop production and quality: First evidence from increased plant biomass, fruit yield and phytomedicine content in bitter melon (*Momordica charantia*). *BMC Biotechnology, 13*(1), 1–10.
40. Kroll, A., Pillukat, M. H., Hahn, D., & Schnekenburger, J., (2009). Current *in vitro* methods in nanoparticle risk assessment: Limitations and challenges. *European Journal of Pharmaceutics and Biopharmaceutics, 72*(2), 370–377.
41. Kumar, D. L., Mitra, J., & Roopa, S., (2020). Nanoencapsulation of food carotenoids. Chapter 7. In: Dasgupta, N., Ranjan, S. and Lichtfouse, E. (eds.), *Environmental Nanotechnology* (Vol. 3, pp. 203–242). Cham: Springer.
42. Kumar, D. L., & Sarkar, P., (2017). Nanoemulsions for nutrient delivery in food. Chapter 4. In: Ranjan, S., Dasgupta, N., & Lichtfouse, E., (eds.), *Nanoscience in Food and Agriculture* (Vol. 5, pp. 81–121). Cham: Springer.
43. Lahiani, M. H., Dervishi, E., Chen, J., Nima, Z., Gaume, A., Biris, A. S., & Khodakovskaya, M. V., (2013). Impact of carbon nanotube exposure to seeds of valuable crops. *ACS Applied Materials & Interfaces, 5*(16), 7965–7973.
44. Lohith, K. D. H., Ankush, S. M., Jagan, M. R. L., & Sowbhagya, H. B., (2017). Microwave impact on the flavor compounds of cinnamon bark (*Cinnamomum Cassia*) volatile oil and polyphenol extraction. *Current Microwave Chemistry, 4*(2), 115–121.
45. Lohith, K. D. H., & Sarkar, P., (2018). Encapsulation of bioactive compounds using nanoemulsions. *Environmental Chemistry Letters, 16*(1), 59–70.
46. Luo, Y., Teng, Z., & Wang, Q., (2012). Development of zein nanoparticles coated with carboxymethyl chitosan for encapsulation and controlled release of vitamin D3. *Journal of Agricultural and Food Chemistry, 60*(3), 836–843.
47. Martins, J. T., Santos, S. F., Bourbon, A. I., Pinheiro, A. C., González-Fernández, Á., Pastrana, L. M., Cerqueira, M. A., & Vicente, A. A., (2016). Lactoferrin-based nanoparticles as a vehicle for iron in food applications - development and release profile. *Food Research International, 90*, 16–24.
48. McClements, D. J., Decker, E. A., Park, Y., & Weiss, J., (2009). Structural design principles for delivery of bioactive components in nutraceuticals and functional foods. *Critical Reviews in Food Science and Nutrition, 49*(6), 577–606.
49. Michelson, E. S., & Rejeski, D., (2006). Falling through the cracks? Public perception, risk, and the oversight of emerging nanotechnologies. *IEEE International Symposium on Technology and Society, 2006*, 1–17.
50. Mushtaq, Y. K., (2011). Effect of nanoscale Fe_3O_4, TiO_2 and carbon particles on cucumber seed germination. *Journal of Environmental Science and Health, Part A, 46*(14), 1732–1735.
51. Nanoproject, [Internet]. *The Project on Emerging Nanotechnologies*. https://www.wilsoncenter.org/publication-series/project-emerging-nanotechnologies (accessed on 14 October 2021).
52. Neethirajan, S., & Jayas, D. S., (2011). Nanotechnology for the food and bioprocessing industries. *Food and Bioprocess Technology, 4*(1), 39–47.

53. Neves, M. A., Hashemi, J., Yoshino, T., Uemura, K., & Nakajima, M., (2016). Development and characterization of chitosan-nano clay composite films for enhanced gas barrier and mechanical properties. *Journal of Food Science and Nutrition, 2*(7), 140–151.

54. Parveen, A., Mazhari, B. B. Z., & Rao, S., (2016). Impact of bio-nanogold on seed germination and seedling growth in *Pennisetum glaucum*. *Enzyme And Microbial Technology, 95*, 107–111.

55. Petersen, K., Nielsen, P. V., Bertelsen, G., Lawther, M., Olsen, M. B., Nilsson, N. H., & Mortensen, G., (1999). Potential of biobased materials for food packaging. *Trends in Food Science & Technology, 10*(2), 52–68.

56. Pinto, T. D. S., Alves, L. A., De Azevedo, C. G., Munhoz, V. H., Verly, R. M., Pereira, F. V., & De Mesquita, J. P., (2016). Layer-by-layer self-assembly for carbon dots/chitosan-based multilayer: Morphology, thickness and molecular interactions. *Materials Chemistry and Physics, 186*, 81–89.

57. Priester, J. H., Ge, Y., Mielke, R. E., Horst, A. M., Moritz, S. C., Espinosa, K., Gelb, Jet al., (2012). Soybean susceptibility to manufactured nanomaterials with evidence for food quality and soil fertility interruption. *Proceedings of the National Academy of Sciences, 109*(37), E2451–E2456.

58. Raman, M., & Doble, M., (2019). Polyphenol nanoformulations for cancer therapy: Role of milk components. In: Huligowda, L. K. D., Goyal, M. R., & Suleria, H. A. R., (eds.), *Nanotechnology Applications in Dairy Science: Packaging, Processing, Preservation* (pp. 123–132). Oakville - ON: Apple Academic Press Inc.

59. Ravanfar, R., Tamaddon, A. M., Niakousari, M., & Moein, M. R., (2016). Preservation of anthocyanins in solid lipid nanoparticles: Optimization of a microemulsion dilution method using the placket-Burman and box-Behnken designs. *Food Chemistry, 199*, 573–580.

60. Rhim, J. W., Park, H. M., & Ha, C. S., (2013). Bio-nanocomposites for food packaging applications. *Progress in Polymer Science, 38*(10), 1629–1652.

61. Sarkar, P., Lohith, K. D. H., Dhumal, C., Panigrahi, S. S., & Choudhary, R., (2015). Traditional and ayurvedic foods of Indian origin. *Journal of Ethnic Foods, 2*(3), 97–109.

62. Savolainen, K., Alenius, H., Norppa, H., Pylkkänen, L., Tuomi, T., & Kasper, G., (2010). Risk assessment of engineered nanomaterials and nanotechnologies: A review. *Toxicology, 269*(2), 92–104.

63. Scrinis, G., & Lyons, K., (2007). The emerging nano-corporate paradigm: Nanotechnology and the transformation of nature, food and agri-food systems. *International Journal of Sociology of Agriculture and Food, 15*(2), 22–44.

64. Servin, A., Elmer, W., Mukherjee, A., De La Torre-Roche, R., Hamdi, H., White, J. C., Bindraban, P., & Dimkpa, C., (2015). A review of the use of engineered nanomaterials to suppress plant disease and enhance crop yield. *Journal of Nanoparticle Research, 17*(2), 1–21.

65. Servin, A. D., & White, J. C., (2016). Nanotechnology in agriculture: Next steps for understanding engineered nanoparticle exposure and risk. *Nano Impact, 1*, 9–12.

66. Shivaram, S. H., & Saini, R., (2019). Spray drying-assisted fabrication of passive nanostructures: From milk protein. Chapter 3. In: Lohith K. D. H., Goyal, M. R., &

Suleria, H. A. R., (eds.), *Nanotechnology Applications in Dairy Science: Packaging, Processing* (pp. 45–68). Oakville - ON: Apple Academic Press.

67. Siddiqui, M. H., & Al-Whaibi, M. H., (2014). Role of nano-SiO$_2$ in germination of tomato (*Lycopersicum esculentum* seeds Mill.). *Saudi Journal of Biological Sciences, 21*(1), 13–17.

68. Siegrist, M., Stampfli, N., Kastenholz, H., & Keller, C., (2008). Perceived risks and perceived benefits of different nanotechnology foods and nanotechnology food packaging. *Appetite, 51*(2), 283–290.

69. Sozer, N., & Kokini, J. L., (2009). Nanotechnology and its applications in the food sector. *Trends in Biotechnology, 27*(2), 82–89.

70. Stampfli, N., Siegrist, M., & Kastenholz, H., (2010). Acceptance of nanotechnology in food and food packaging: A path model analysis. *Journal of Risk Research, 13*(3), 353–365.

71. Sultana, K., Godward, G., Reynolds, N., Arumugaswamy, R., Peiris, P., & Kailasapathy, K., (2000). Encapsulation of probiotic bacteria with alginate-starch and evaluation of survival in simulated gastrointestinal conditions and in yoghurt. *International Journal of Food Microbiology, 62*(1), 47–55.

72. Tsai, P. J., Chen, Y. S., Sheu, C. H., & Chen, C. Y., (2011). Effect of nanogrinding on the pigment and bioactivity of djulis (*Chenopodium formosanum* koidz.). *Journal of Agricultural and Food Chemistry, 59*(5), 1814–1820.

73. Wan, Z., Wang, L., Yang, X., Guo, J., & Yin, S., (2016). Enhanced water resistance properties of bacterial cellulose multilayer films by incorporating interlayers of electrospun zein fibers. *Food Hydrocolloids, 61,* 269–276.

74. Wijnhoven, S. W., Peijnenburg, W. J., Herberts, C. A., Hagens, W. I., Oomen, A. G., Heugens, E. H., Roszek, B., et al., (2009). Nano-silver: A review of available data and knowledge gaps in human and environmental risk assessment. *Nanotoxicology, 3*(2), 109–138.

75. Zou, L., Zheng, B., Zhang, R., Zhang, Z., Liu, W., Liu, C., Xiao, H., & McClements, D. J., (2016). Food matrix effects on nutraceutical bioavailability: Impact of protein on curcumin bioaccessibility and transformation in nanoemulsion delivery systems and excipient nanoemulsions. *Food Biophysics, 11,* 142–153.

CHAPTER 2

SCOPE AND APPLICATIONS OF NANOTECHNOLOGY IN HORTICULTURE

VIJAY S. RAKESH REDDY, GAJANAN GUNDEWADI, and
LOHITH KUMAR DASARAHALLI-HULIGOWDA

ABSTRACT

Being in the nascent stages of development, nanotechnology has a wide
spectrum of applications, both on-farms and off-farm in horticulture
especially in the developing countries. It is expected to reduce the
post-harvest losses, improve the product quality, and simultaneously
increase the competitiveness of the horticultural products. The science of
nanotechnology could also be effectively applied in the development of
new functional materials, products; design of methods and instrumentation
for enhancing the productivity, bio-security, and food safety. Under such
situations, our farmers need to be enlightened and carefully guided
regarding the pros and cons of nanomaterials.

2.1 INTRODUCTION

Nanotechnology is a multidisciplinary field, which comprises of research,
technology development, and control of structures within the range of
nano-scale (1 to 100 nm). It includes manipulation or self-assembly of
discrete atoms, molecules, or their gatherings for creation of devices
and materials possessing unique properties. This technology has been
developed as result of efforts from the researchers and engineers in
photonics, microelectronics, biotechnology, and chemistry. It majorly

involves the emerging applications of nanoscience (viz.: characterization, fabrication, and manipulation of substances) at nano-scale. When the particle size of a substance is compacted under its threshold level, the consequential material would exhibit significantly different properties (physical and chemical) from those of the macroscale substances.

The nano-sized particles with greater surface area per unit mass have proven to be biologically more active compared to their respective larger sized particles of similar chemistry [79]. With reduction in particle size, this technology could also improve the water solubility, thermal stability, and overall bioavailability of functional composites through improvement of their delivery properties, solubility, extended residence time in the gastro intestinal tract, and proficient absorption through cells [38, 58]. A nanomaterial is any substance containing one or more dimensions in the nano-scale range; however, a nanoparticle is a discrete entity with all the three dimensions in the range of nano-scale. These nanomaterials and nanoparticles (NPs) can be composed into nanofilms, nanosheets, nanolayers, nanotubes, nanorods, nanofibers, nanocoating, etc. [20].

Nanotechnology can be applied effectively in the design and development of processes and instrumentation for enhancing quantity and quality of horticultural produce with better management of the environment [23]. However, the implication of nanotechnology in the horticulture sector can be far reaching. The nanomaterials occur naturally in most of the plant and animal products, such as: major constituents of milk, fibrous structures of fish and meat, crystalline structures of innate starches, and cellulose fibrils of plants [65, 74]. Synthetically engineered nanomaterials are also available for a variety of horticultural applications.

This chapter focuses on the aspects of nanotechnology that have direct implications in the crop improvement, production, protection, post-harvest management and processing of various horticultural crops. In addition, authors have also discussed the identification of current problematic areas in the application of nanotechnology to horticulture in view of their potential risk to human health and environment, regulatory issues, and public discernment.

2.2 NANOTECHNOLOGY: SCOPE AND IMPORTANCE IN HORTICULTURE

Being a multidisciplinary science, nanotechnology has enormous potential to boost up the horticultural research. Nanotechnological developments have resulted from the integration of biology, chemistry, photonics, food science, biotechnology, and electronic engineering. For example, nanofabrication involves integration of Biology with Material Science and Engineering [81]. Horticulture is a vital part of the wider biological industry. Since, the whole biological world resides within the sphere of nanotechnological scale, there is a strong logic in convergence of Nanobiology, Biotechnology, and Bioengineering to solve the practical problems faced by the horticultural industry.

Recently, the novel materials and surface characteristics resulting from the application of this nanoscience could be utilized for enhancing horticultural production and productivity. This technology can also be used as a tool in better understanding of the various cellular processes, mechanisms regulating major horticultural traits, and in development of genotypes tolerant to biotic and abiotic stresses. Thus, nanotechnology could offer better products and improved means of production in near future. Based on the current research advances in the field of nanotechnology, it could completely transform the horticultural industry by altering the way of production, processing, packing, transportation, and consumption.

If the horticulture is to attain its broad goal of sustainability, then this novel technology should be extended across the entire horticultural value chain. In horticulture, the nanotechnology has wider applications in (Figure 2.1) crop improvement, soil management, plant disease diagnostics, efficient agrochemical delivery, water management, bioprocessing, postharvest management, processing, packaging, monitoring the identity and quality of horticultural produce, and precision horticulture [116].

Other potential applications of nanotechnology in horticultural systems include: nanocoating, nucleic acid bio-engineering, bio-analytical nanosensors, and bio-selective surfaces, etc. Development of low-cost nanosensors has been possible for detection of food borne pathogens. Micronutrient deficiency and mineral toxicity was found to be the major limitation in the horticultural production. High input use efficiency can be attained through use of nano carriers for smart delivery to the targeted sites and nanoformulations for controlled chemical release. Nanosensors can be utilized to study the soil nutrient status and the soil microflora. Use of systemic herbicides against parasitic weeds in the form of nanocapsules could avoid the phytotoxicity of the crop.

Since nanotechnology is manipulated atom by atom, the processes and products developed from them are most precise, which were literally impossible to produce through the conventional methods and processes. Thus, nanotechnology is envisaged as a rapidly emerging science with a great potential to modernize horticulture. This has changed the research priorities in most of the developed and developing nations. However, till date, >90% of the nano-based patents and products were developed in USA, China, Germany, France, Japan, Switzerland, and South Korea.

In recent years, there has been an exponential increase on both investment and development of nano-based processes and products. The natural resources (such as: water, nutrients, and fertilizers) are to be used judiciously and efficiently in the horticulture systems and this is best possible through nanosensors and other nano-based smart delivery systems. The quality of the horticultural produce can be monitored through nano-barcodes and other nano-based tracking systems. Thus, the infusion of nanotechnology into horticultural science would transform the conventional farming practices to precision horticulture to ensure food security [105, 106]. This includes early prediction of the biotic stresses followed by their management, enhancement of input use efficiencies, and shelf-life of the perishable products (viz.: fruits, vegetables, and flowers).

2.3 APPLICATIONS OF NANOTECHNOLOGY IN HORTICULTURE

2.3.1 NANOTECHNOLOGY FOR CROP IMPROVEMENT

Nanotechnology has the potential to improve horticultural productivity through genetic improvement of crops, delivery of genes and drug molecules to targeted sites at the cellular level, and nano-array based gene-technologies for gene expressions in crop plants under stress conditions. All these are made possible due to various appealing features of DNA (deoxyribonucleic acid) molecule, viz.: its miniscule size (with diameter of 2 nm), its short structural repeat (helical pitch of 3.4–3.6 nm), and its stiffness (with persistence length of 50 nm). The DNA molecule acts as a key player in the bottom-up nanotechnology as its assembly system was completely chemical (bonds). This approach originated in the early 1970s, where *in vitro* genetic manipulations were carried out using "sticky ends" for tacking together of different DNA molecules (Figure 2.1).

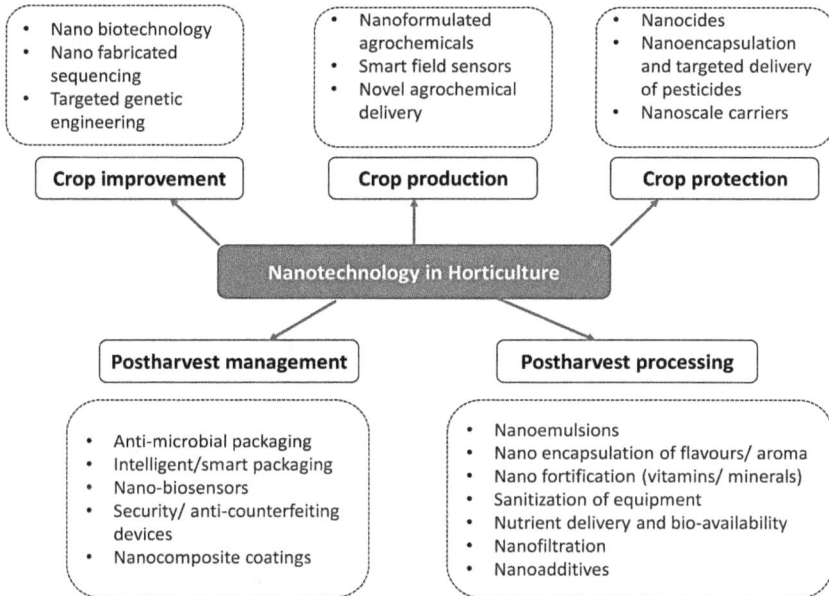

FIGURE 2.1   Nanotechnological applications in horticulture science.

The sticky ends with complementary nucleotide bases would cohere to form a molecular complex. The sticky end cohesion is probably the finest example for programmable molecular recognition. In a similar way, the solid-support based DNA synthesis makes it easy to cohere diverse sequenced sticky ends. Thus, this sticky end technology provides both predictable controls of intermolecular associations along with the geometry of cohesion. Although similar affinity properties can be achieved through cohesion of specific antibodies and antigens, yet they may need standardization of their relative three-dimensional orientations for each new pair. Thus, the nucleic acids were inimitable in this regard, as they provide a tractable, diverse, and programmable system.

Further, developments of nanoscience in this sector could become the main driving forces for economy in the long-run benefitting the producers, farmers, consumers, ecosystems, and the society in general. Though this technology could raise few novels, social, ethical, philosophical, and legal issues, yet it can be predicted and alleviated through appropriate regulatory mechanisms.

2.3.1.1   NANOBIOTECHNOLOGY

The nanobiotechnology acts as a link between the breakthroughs of nanotechnology and the molecular biology. The molecular biologists play major role in helping the nanotechnologists to understand and access the nanomachines and structures designed through the process of evolution. Many challenging goals can be accomplished easily by the nanotechnologists through exploitation of the extraordinary properties of the biological molecules and cell processes. For instance, the DNA ladder structure offers a natural framework for assembling nanostructures instead of building a silicon scaffolding; also, the highly specific bonding properties of DNA molecule can bring the atoms for creation of desired nanostructure. Some nanotechnologists rely frequently on self-assembling properties of the biomolecules for creating certain nanostructures, such as, spontaneous formation of liquid crystals from lipid molecules [113].

In addition to the building of nanostructures, DNA is also useful in making of nanomachines. With recent advancements in information technology, the DNA being data storage molecule could serve as basis for subsequent generation computers. With the shrinkage of microprocessors and microcircuits to nanoprocessors and nanocircuits, the silicon chips mounted with DNA molecules could be the traditional microchips etched with electron flow channels. These biochips with DNA-based processors could use the DNA's bizarre data storage capacity.

The nanomaterials could also be used as effective delivery modes for transportation of nucleotides and other specific chemical molecules into plant cells [77]. Researchers have tried to use honeycomb mesoporous silica NPs (3 nm pore size) for delivery of nucleic acids and other chemicals into the intact leaves and plant cells. They have loaded the desired genes along with their chemical inducers into the mesoporous silica NPs and capped their ends with gold NPs. Later, they observed the release pattern of various chemicals along with the initiation of gene expression in those plants under controlled conditions. All this has indicated that the silica NPs can be selectively used for target specific delivery of the proteins, nucleotides, and associated chemicals in the plant nanobiotechnology [108]. Various modes of plant delivery systems are indicated in Table 2.1.

Nanobiotechnology fuses nano/microfabrication and biosystems. This relates to all kinds of genomic applications comprising of plants, mammals, and microbes. It offers simple tools and technology for assembling sequence

TABLE 2.1 Applications of Nanobiotechnology for Gene Transfer in Plants

Nanomaterial	Application	Function	References
Fluorescence starch-nanoparticle coated with poly-L-lysine	As plant transgenic vehicle	DNA nanoparticle complexes bind and transport genes across the cell walls by induction of pore channels in them with the help of ultrasound treatment.	[60]
Gold functionalized mesoporous silica nanoparticles (Au-MSN)	Co-delivery of DNA and plasmid into plant cells through biolistic method	Protein loaded Au-MSN (10 nm pore) were coated with plasmid DNA and delivered into plant tissues through particle bombardment.	[68]
Honeycomb mesoporous silica nanoparticles	Delivery of DNA and other molecules	The mesoporous silica nanoparticles (3 nm pores) loaded with a gene and its specific chemical inducer could be successfully delivered into the plant cells and trigger their expression under controlled release conditions.	[108]
Magnetic carbon coated nanoparticles	Absorption and translocation through the root to different plant parts	In peas and tomato, the nanoparticles get penetrated through the roots, reach the vascular bundles through transpiration, and spread to aerial parts of the plant within 24 hours	[18]
Plant biotransformable HMG-CoA reductase gene loaded with calcium phosphate nanoparticle (Cap)	Encapsulation of plasmid DNA into nanoparticle for enhancing the stability and frequency of genetic transformation in plants	In acidic media, the DNA release was initially slow followed by fast release of Cap nanoparticles. Though stable at ambient temperature and RH, it significantly enhanced the transformation efficiency along with the antioxidant capacity of the transformed plants.	[91]
Single-walled carbon nanotubes	Transporters in plant cell walls	They penetrate the cell wall and cell membrane of intact plant cells and deliver different cargoes into various cell organelles successfully	[61]
ZnS nanoparticles modified with positively charged poly-L-lysine	Delivering DNA into cell	Successful delivery of β-glucuronidase encoding plasmid DNA into plant cells by means of ultrasound assisted methods.	[119]

data and application of this information in medicine, agriculture, and horticulture.

Potential applications of nanobiotechnology in horticulture include: high throughput DNA sequencing; nanofabricated gel-free systems; microarray and gene expression profiling; rapid, powerful, and accurate diagnosis of plant diseases; scaling down of biosensors; creation of bio-nanostructures for delivery of functional molecules into cells, etc.

2.3.1.2   NANOFABRICATED DNA SEQUENCING

As DNA sequencing is the central process, therefore, it needs to be upgraded in terms of accuracy and throughput. Nanofabrication technology is highly critical for improving the existing conventional methods and to develop novel approaches of DNA sequence detection. Miniaturizing the existing sequencing technology will permit the process to be more analogous and multiplex.

The research in nanobiotechnology is progressing towards the sequencing of the DNA molecules with rapid nanofabricated gel-free systems. This nanofabricated sequencing technology along with powerful techniques (viz.: association genetic analysis, DNA sequence data of crop germplasm (crop gene pool along with their wild relatives)) could possibly deliver vital information on molecular markers associated with agronomic, horticultural, and economic traits. Thus, nanobiotechnology could improve marker-assisted breeding for horticultural crop improvement.

2.3.1.3   MICROARRAY AND EXPRESSION PROFILING

Microarrays and expression profiling have become principal tools in the biological research system, wherein they measure the expression levels for thousands of genes simultaneously that could contain the information regarding various aspects of gene regulation and function. Hence, there is a great need for development of novel formats sequence and gene expression patterns with higher throughput than the current technologies. Earlier, thousands of DNA or protein molecules organized on glass slides were used for creation of DNA and protein chips, respectively; but with the advancement of the microarray technology, they use custom-made beads in place of glass slides. Ultimately nanofabrication technology can be used to pattern surface chemistry for a range of biosensors and biomedical applications. This microarray technology can be extensively used in determination of new genomic sequences, gene scanning for polymorphism with phenotypic impact, and for the comprehensive appraisal of the array of gene expression in plants when subjected to biotic and abiotic stresses. With the incorporation of nanoscience into microarray technology, the researchers are creating many types of microarrays, such as: DNA microarrays and protein microarrays for detection of mutation in disease related genes; monitoring of gene activity; and identification of vital genes for crop productivity, etc.

2.3.2 NANOTECHNOLOGY FOR CROP PRODUCTION

Nanotechnology plays a significant role in the promotion of sustainable horticulture for providing better food. In the developing countries, this technology has got significant applications for enhancement of horticultural productivity, along with biotechnology, genetics, plant breeding, disease control, nutrient application, precision horticulture, and all other related fields [46, 92]. The vital data relating to crop growth and precision farming practices can be availed on real-time basis using the nanosensors and other field sensing nano devices, which could lead to input minimization and maximization of resource output, i.e., yield [13, 95]. These nanosensors also provide information on crop nutrient status, optimal planting and harvesting periods, and application timing of various agrochemicals.

2.3.2.1 NANOPARTICLES (NPS) FOR PLANT GROWTH ENHANCEMENT

Nanomaterials can be used efficiently in the seed germination and growth of horticultural crops. Among different nanostructures, carbon nanotubes (CNTs) have been used as regulators of seed germination and plant growth. It was found that multi-walled CNTs possess the ability to boost up the growth of cell cultures by 55–64% at concentration of 5 to 500 µg/ml. The lower concentrations of activated carbon could promote cell-growth, while higher concentrations were inhibitive in nature [50]. In the process of increasing the seed germination rate, the permeation of nanomaterial into the seed coat could play an important role. Likewise, the CNTs are found to regulate the cell division and growth through a unique mechanism of activating the aquaporins (water channels). However, to extract the predicted benefits of the carbon NPs in horticulture, there is a great need to investigate the ultimate effects of the CNTs on the environment.

Khodakovskaya et al. [50] have shown that the increased germination percentage (90%) of the tomato seeds treated with CNTs compared to the control (71%) might be due to the penetration of CNTs into the hard seed coat of tomato seeds. Additionally, these multi-walled CNTs could be efficiently used to deliver certain preferred molecules into the seeds during germination and shield them from diseases. As they have growth promoting effects, they will not cause any toxic/ adverse effects.

On the other side, single-walled CNTs were functionalized with poly-3-aminobenzenesulfonic acid; and its influence on the root elongation of six economically important vegetable crops (cabbage, lettuce, cucumber, tomato, onion, and carrot) was studied [8]. Also, the non-functionalized single-walled CNTs affected the root growth significantly compared to their functionalized counter parts.

The scanning electron microscopy (SEM) has shown the presence of nanotube sheets on the root surface with no perceptible uptake of nanotubes by the plant. Likewise, the effect of nano and non-nano TiO_2 (Titanium dioxide) on the development of spinach seeds was observed by Zheng et al. [121], who reported that the plants produced from the nano TiO_2 treated seeds have accumulated 73% more dry weight with increased chlorophyll-A formation leading to three-fold rate of photosynthesis compared to the untreated samples. It was also reported that the toxicity, behavior, and reactivity of the nanomaterials is mainly controlled by their particle size and the percentage of germination increases with decrease in the particle size. Thus, appropriate investigation on the hostile effects of the nanomaterials on the seed and plant properties can increase their effective utility in the horticultural crops.

2.3.2.2   NANOTECHNOLOGY IN WATER MANAGEMENT

There is a great potential for use of nanomaterials for the treatment of surface and ground waters in addition to the waste water (viz.: industrial effluent water, gray water from the urban dwellings and other water contaminated by the toxic metal ions, organic, and inorganic solutes along with certain harmful microorganisms). Because of their distinctive activity towards recalcitrant impurities, numerous nanomaterials are under trail for use in water purification. For production of healthy crops without any heavy metal contaminants/ microbial inoculum, the irrigation water should be analyzed on a real time basis. Unfortunately, the traditional laboratory tests are time consuming and have meager repeatability. Hence, rapid procedures involving enzymes, immunological/genetic tests are being developed.

The water quality could be significantly enhanced with the use of nano-fiber membranes for filtration and nanobiocides for killing the microbial contaminants. Biofilms contaminating potable water are actually bacterial mats enveloped in natural polymers and are hard to treat with antimicrobials or other chemicals. Although they could be cleaned mechanically, yet they cost considerable labor and down-time. Hence, there is a need to develop nanoenzyme treatment to breakdown such biofilms effectively [90].

2.3.2.3   NANOFERTILIZERS FOR BALANCED NUTRITION

In general, application of specific fertilizers at appropriate stage of crop growth could increase the productivity to the tune of 35–40%. However, its overuse has resulted: (1) in imbalanced fertilization and nitrate pollution of ground water in certain areas; and (2) in accumulation of fertilizers in soil and eutrophication of water bodies though the leaching effect. Hence, there is a need to develop new and novel nano-based fertilizers to address the issues relating to low efficiency, imbalanced applications, micro-nutrient deficiencies and declining of soil organic matter.

Though there is a great scope for the formulation of nano-fertilizers, yet the innovations are scanty. Tarafdar et al. [106] observed substantial rise in the crop yields with foliar application of the nanoparticle-based fertilizers. Foliar application of nanophosphorus (@ 640 mg/ha) has increased the yield (80 kg/ha) of cluster beans under arid environment. The bulk calcium is immobile and its mobility in the phloem tissues can be improved with application of nano-based Ca [21].

Currently, the research is under a big way to develop nanocomposites that could supply all vital nutrients in appropriate quantities through smart delivery systems. However, the metabolic assimilation of micronutrients within the plant biomass applied as nanoformulations through soil/ foliar or other means should be ascertained. Further, the nitrogen supplied through conventional fertilizers has low use efficiency due to 50–70% losses.

Therefore, there is a need to exploit the porous nanoscale plant parts for improving the plant uptake and reducing the nitrogen losses. Fertilizers can also be encapsulated with NPs and their release can be triggered by environmental conditions or at pre-desired intervals.

2.3.3   NANOTECHNOLOGY FOR CROP PROTECTION

Application of nanoscience for protection of horticultural crops holds a significant promise in the management of weeds, insect pests and pathogens. This is possible with the targeted delivery of agrochemicals along with provision of early detection diagnostic tools. The NPs are extremely stable and are eco-friendly in nature; hence they could easily be used in the smart conveyance of agrochemicals (such as: herbicides, pesticides, and fertilizers) through their nanocapsules. The nanoscale carriers can increase the effectiveness through firm binding and ultimately preventing the runoff of harmful agrochemicals into groundwater [13, 47]. Since the encapsulated NPs exhibit slow release of functional molecules, they reduce their frequency of application. The NPs behave differently from their bulky parts, since their nanosize enables them with greater solubility and stability due to more surface area and greater charge. The effectiveness of oxide-based NPs (viz.: zinc oxide (ZnO), aluminum oxide, TiO_2) and AgNPs against insect pests and pathogens were studied by Goswami et al. [37].

Nanosilica showed 100% insect pest mortality, while nanosulfur compounds inhibited the sporulation and growth of fungi. In the green synthesis approach, the biological agents (viz.: plants and microbes) could be used as efficient and cost-effective candidates and have special advantage over the conventional chemical methods with reduced production of harmful waste through use of eco-safe solvents and non-hazardous compounds [99]. Due to low viscosity, high kinetic stability and optical transparency, NPs are best suited as smart carriers for various applications. Ultimately these NPs are feasible substitutes for the traditional pesticides, for which some pests have developed resistance.

2.3.3.1   WEED MANAGEMENT

Most of the commercial weedicides destroy/ terminate the aerial plant parts and none of them are able to inhibit the activity of the underground parts (such as: rhizomes and tubers) that act as source of seed for the subsequent seasons. In general, low harvests are obtained from the weed and weed seed-ridden soils compared to those free from weed infestation. Development of new efficient herbicides using nanotechnology might result in better production of horticultural crops.

The encapsulated nanoherbicides are highly pertinent to the rainfed horticulture, as they truly mimic the rain-fed system, where they are well-protected under normal atmospheric conditions and get activated only with the spell of rainfall. Designing and production of target specific herbicide molecules encapsulated with NPs and targeted at particular receptors in the roots of specific weeds, wherein they enter the root system, get trans-located to various plant parts and inhibit the glycolysis of the food reserves in the plant. This would ultimately starve and kill the weed plant for lack of nutrients [16].

Currently, the adjuvants containing nanomaterials for herbicide application are available in the market. For example: Use of soybean micelle-based nano-surfactant has made the glyphosate-resistant crops susceptible to glyphosate. Silva et al. [102] showed that association of the herbicide Paraquat with the chitosan-based NPs (635 nm) has resulted in significant difference of the release of profiles compared with the free Paraquat. This kind of loading herbicides into NPs have considerable impact in reducing the negative impact of harmful synthetic agrochemicals.

2.3.3.2 PEST MANAGEMENT

The insect pests are reported to cause colossal crop losses to the extent of 14% [85]. During the last decade, the average pesticide production was about three million tons; and out of this only very small amount (<0.1%) reaches the site of action due to losses during application in the form of spray drift, run-off, off-target deposition, and phyto-degradation [9]. These pesticides can cause adverse effects on human health and the pollinating insects. Since there is a growing demand for pesticide usage globally, therefore there is an urgent need to develop nanoformulated pesticide with site specific mode of action and eco-friendly in nature. Hence, nano-technology plays a vital role in development of nanocides with decreased toxicity and increased efficacy [76].

The bio-availability of the poorly water-soluble pesticides could be improved with nanoparticle additives or the NPs could be laden with pesticides and programed to release gently based on the environmental trigger [49, 57]. Also, nanoencapsulation of agrochemicals for slow release to a specific host plant for controlling specific insect pest would allow the active ingredient to be appropriately absorbed by the plants [96].

The utility of some water insoluble botanicals (such as: rotenone) is restricted due to their meager water solubility, low stability, degradation, and isomerization on sunlight exposure [55]. Synthesized NPs loaded with the insecticide rotenone were 1,300 times more effective compared to the free rotenone in water. Researchers have found that powder formulations of NPs are more effective in controlling the insect pests. In addition to this, various nanoformulations were developed to improve the efficiency of the agrochemicals (such as: microemulsions, nanoemulsion, nanodispersion, polymer-based (Table 2.2) and solid-lipid based NPs, porous hollow silica-based NPs, double layered hydroxides, and metal-based NPs for the crop safeguard [49].

Lasic [56] found that encapsulation of biocides into liposomes might prolong the action of these nanocides on crop plants by avoiding surface wash-off and also help in effective delivery of essential nutrients to the plants. Wang et al. [111] established the latent use of water insoluble pesticide (i.e.: β-cypermethrin) as nanoemulsion to increase the stability of its spray solution. Also, there are few reports on the development and application of nanosilica as a nanopesticide [4].

2.3.3.3   DISEASE MANAGEMENT

The horticultural crops are confronted by diverse groups of bacterial and fungal pathogens. Some of the common bacterial agents belong to the genera, *viz.*, *Erwinia*, *Pseudomonas*, *Corynebacterium*, *Xanthomonas*, etc.; while the fungal agents belong to the genera, such as: *Alternaria*, *Aspergillus*, *Colletotrichum*, *Cladosporium*, *Fusarium*, *Penicillium*, *Pythium*, *Phytopthora*, *Rhizobium*, *Botrytis*, etc. Though some of them are host specific, yet most of them have broad host range and thus may cause massive financial losses. Few of these may produce toxic metabolites and adversely affect the consumer health. Most of them can gain entry into the plant tissue through mechanical or chilling injury and cause devastating losses [109]. Nanotechnological application in plant disease management targets specific horticultural problems by studying the plant-pathogen interfaces and providing new insights for crop protection (Table 2.3). Nano-composites justify the two most important criteria in disease management:

(i) greater effectiveness with negligible impact on environment; and (ii) less toxicity to the humans.

TABLE 2.2 Some Nanopesticide Formulations Made Using Different Polymers

Active Ingredient	Nanomaterial	Polymer Used	References
Imidacloprid	Capsule, granules	Lignin-polyethylene glycol-ethyl cellulose	[32]
	Particle	Chitosan-poly (lactide)	[59]
	Clay	Bentonite	[31]
Novaluron	Powder	Anionic surfactants viz. sodium dodecyl sulfate, naphthalene novaluron sulfonate condensate sodium salt, etc.	[25]
Carbofuran	Suspension	Poly (methyl methacrylate)-poly (ethylene glycol)-polyvinyl pyrrolidone	[15]
	Micelle	Polyethylene glycol-dimethyl esters	[98]
Pheromones	Resin	Vinylethylene and vinylacetate	[115]
	Fiber	Polyamide	[39]
Aldicarb	Gel	Lignin	[52]
Azadirachtin	Particle	Carboxymethyl chitosan-ricinoleic acid	[29]
Carbaryl	Capsule	Carboxymethylcellulose	[43]
Chloropyriphos	Particle	Polyvinylchloride	[62]
Endosulfan	Film	Starch-based polyethylene	[44]
Garlic essential oil	Capsule	Polyethyleneglycol	[118]
Neem seed oil	Capsule	Alginate-glutaraldehyde	[1]
Piperonyl butoxide and deltamethrin	Capsule	Polyethylene	[33]

2.3.4 NANOTECHNOLOGY FOR POSTHARVEST MANAGEMENT

We are not able to utilize all horticultural produce produced by our farmers due to greater perishable nature of the horticultural produce. The best solution for this problem could be use of nanotechnology and biotechnology for increasing the production efficiency and decreasing the postharvest wastage.

TABLE 2.3 Efficacy of Nanomaterials in Controlling Diseases of Various Horticultural Crops

Crop	Type of Nanoparticle	Action Against	Mode of Action	Function and Applications	References
Cabbage	Silver NP	*Xanthomonas campestris* pv. *campestris*	They have strong antibacterial activity and exhibit higher toxicity to microbes than to the mammalian cells	• The antibacterial activity of the nanoparticles significantly reduced the cabbage black rot • The disease reduction was in a dose dependent manner	[34]
Chilly	Chitosan NP	*Rhizopus sp.; Colletotrichum capsici; C. gloeosporioides;* and *Aspergillus niger*	The positively charged chitosan molecules may interact with the negatively charged phospholipid components of microbial membranes thus altering their permeability and causing cellular leakage; It may also chelate with the metal ions and thus effect microbial growth; it can also penetrate the fungal cell wall and bind with its DNA inhibiting mRNA production.	Significantly delayed the mycelial growth at a concentration of 0.6% (w/v) and improved the chili seed quality	[17]
Pepper	Thiamine Di-lauryl sulfide (TDS) nanoparticles	*Colletotrichum gloeosporioides*	TDS nanoparticles have antifungal activity as they penetrate the hyphal cell membrane and destroy the cell membranes	The inhibition was around 80% when the particles were used in the concentration range of 100 ppm	[97]
Pome-granate	Nano copper	*X. axonopodis* pv. *punicae*	Production of highly reactive hydroxyl radicals that could damage proteins, lipids, DNA, and other biomolecules	Inhibited the growth of blight bacteria at 0.2 ppm concentration, i.e., 10,000 times lower than the recommended copper oxychloride	[70]

TABLE 2.3 *(Continued)*

Crop	Type of Nanoparticle	Action Against	Mode of Action	Function and Applications	References
Rose	Light-activated NP formulation of TiO₂ with Zn	Leaf spot bacteria	Production of hydroxyl and superoxide radicals mediated the cell wall deformity and caused death by high energy transfer	The nano formulations at a concentration of 500–800 ppm in the field has significantly reduced the bacterial spot severity	[82]
Tomato	Light activated nanoscale formulations of TiO₂	*X. perforans*	TiO₂ in nanoformulations with Ag and/ or Zn have greater photocatalytic activity against X. perforans	These composites significantly reduced the severity of bacterial spot in tomato	[83]
Tomato	Copper based nanoparticles	*Phytophthora infestans*	Production of highly reactive hydroxyl radicals that could damage proteins, lipids, DNA, and other biomolecules	• The Cu nanoparticles (11–55 nm) were more effective than the commercial fungicides at lower concentrations. • They do not show any symptoms of phytotoxicity on the treated plants.	[36]
Tomato	DNA directed silver nanoparticles on graphene oxide	*X. perforans*	Strong microbial inhibition by silver ions and exhibition of higher toxicity to microbes and lower toxicity to the mammalian cells	The bacterial spot disease was significantly reduced by 100 ppm silver nanoparticles.	[80]

Nanotechnology helps in extending the shelf-life of the horticultural products through control of growth and development of microbes, introduction of new generation of package films, control of the influence of gasses and harmful UV (ultraviolet) rays, increased strength, quality, and consumer appeal along with use of nano biosensors (multiple chips) for labeling products which was essential in automatic labeling of products [116].

Nanotechnology has great potential application in enhancing the life of cut flowers by regulating the ethylene in storage chamber and during transport. The main aim of nanoscience in this area is to provide good quality flowers to consumers by using innovative techniques [93]. The ethylene management strategy of nanotechnology is shown in Figure 2.2. Nanoscale system could be useful for ethylene detection in storage chamber of cut flowers by using Nano-metal based sensors and nanochip labels along the distribution chain. Most commonly used ethylene gas sensors in detection and monitoring of ethylene gas are: tin dioxide (stannic oxide: SnO_2), tungsten trioxide (WO_3), palladium (Pd), platinum (Pt), titanium dioxide (TiO_2), and ZnO [93].

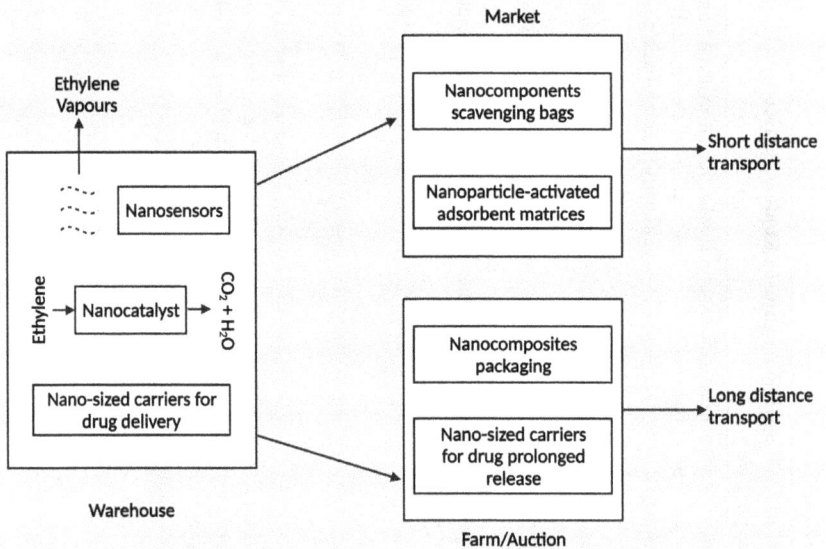

FIGURE 2.2 Ethylene management strategy for ethylene-sensitive cut flowers.

Nanocomposites and nanocatalysts are used for removal and photo degradation of ethylene, i.e., nanocomposites for scrubbing in active packaging and nano-metals for photocatalytic degradation in the store house. From the practical point of view, the best tested catalysts for ethylene removal have been immobilization of Pd and TiO_2 on activated carbon.

Titanium dioxide (TiO_2) has been utilized for photo catalytic degradation, i.e., specifically UV irradiation either from artificial lamps or natural sunlight. However, its low toxicity, low cost, chemical, and physical stability have allowed researchers to achieve functional nanocomposites. Silver ions also show photoactivity, semiconductor photocatalysis and antimicrobial activity; nano-Ag absorbs and destroys ethylene and can have more effective antimicrobial activity than silver (Ag). Thus, packaging films integrating nano-Ag or TiO_2 (e. g. Nanocomposite polythene film) contribute to preserve the quality of fresh fruits and vegetables, by delaying senescence and inhibiting microbial growth [93].

2.3.4.1 NANOPACKAGING

While most of the nanotechnological applications in the food and beverage industry are in the nascent stages of R&D (research and development) or in near-marketing stages, the application in the packaging sector has already become a commercial reality [11]. The nanopackaging application is predicted to have a maximum share of 19% among all the nano-enabled products in the current market. The incorporation of nanomaterials into the plastic polymers has led to the development of various innovative packaging materials [10], such as:

- Active nanocoatings for sterile food contact surfaces;
- Hydrophobic nanocoatings for self-cleaning surfaces;
- Nano biosensors for smart/intelligent packaging systems.
- Polymer nanocomposites with improved properties; and
- Polymers incorporating antimicrobial nanomaterials.

Nanopackaging offers hope for horticultural crops through extending shelf-life, safer packaging, better traceability of horticultural products, and healthier food. Packaging material incorporated with nanocomposites may enhance material properties (such as: strength, thermal stability, gas barrier) and antimicrobial properties, without affecting their toughness and transparency [54]. Most of the horticultural crops are seasonal in nature and to extend their availability over the off-season, nanopackaging is one of the best options to serve numerous functions, such as: protection from moisture, dust, oxygen, light, harmful microorganisms, etc. For effective utilization, the packaging material should also be safe to use, inert, eco-friendly, cost-effective, light weight, easy to dispose, reusable, good stake strength, ability to withstand extreme processing conditions, such as: filling, sealing, and transport conditions.

Fresh horticultural produce (flowers, vegetables, fruits) is metabolically active and needs to be packaged in O_2 permeable materials with an optimal transmission rate, while the processed horticultural products does not require such gaseous transfer. The critical issues relating to packaging material for extending the shelf-life of perishable products is their poor barrier properties to water vapor and other gasses, as the fruits and vegetables respire continuously by taking in oxygen and giving out CO_2.

In case of apple, the oxygen causes apple scab on its surface. Another major critical issue in packaging of processed foods is that of moisture and gaseous migration and permeability: no packaging material is fully resistant to atmospheric gasses, water vapor, or natural substances of the contained food or even the wrapping material itself.

Traditionally jute bags, bamboo baskets, wooden crates, corrugated fiber board (CFB) boxes, etc., are being used for handling fresh horticultural produce; while metal, ceramic (glass), laminated paper, and plastics are used for packaging of processed fruit and vegetable products. Though these materials are still under use, yet plastic based packaging materials are getting more popularity in the current scenario due to their low-cost, light weight, ease of processing and development. Polymers often used in packaging are: polypropylene (PP), high-density, and low-density polyethylene (HDPE, LDPE), polyethylene terephthalate (PET), polystyrene (PS) and polyvinyl chloride (PVC). Although these polymers are considered as the present-day packaging material, yet they have certain major drawbacks, viz., their natural permeability to gasses and other small molecules. Permeability of a polymer to O_2 or moisture is dependent on its polarity, degree of branching, degree of crystallinity, hydrogen bonding characteristics, method of synthesis, molecular weight, polydispersity, processing methodology, and structure of polymeric side chains.

Certain Polymer films are crammed with "silica nanoparticles" to substantially decrease the O_2 in-flow and moisture out-flow to keep the produce fresh and appealing to the consumers. Plastic material for packaging of fruit- and vegetable-based beverages must have high O_2 and CO_2 barrier property to check decarbonation of juices and oxidation of their contents. As a result of these complexities, processed horticultural products require sophisticated packaging material for extending the product shelf-life and prevention of spoilage. New eco-friendly materials have been exploited for the development of edible films to reduce packaging waste [107]. However, the use of edible polymers has been limited due to their brittleness, poor barrier (gas and moisture) properties, low heat distortion temperature and high cost. The application of nanoscience to these polymers may open up new prospects for refining the barrier properties as well as improving the cost-price-efficiency [103].

2.3.4.1.1 Nanocomposites

Packaging material incorporated with nanocomposites may serve as smart packaging by regulating the oxygen levels for enhancing the quality and safety of horticulture products and for interacting with consumer in case of spoilage or pathogen detection [117]. The use of bio-polymers in preparation of novel nanocomposites offer the rare opportunity of carbon-neutral biodegradable films for packing of horticultural produce (Table 2.4). This also offers a great opportunity in developing countries for best utilization of their agricultural and forestry by-products and waste materials for producing low-cost bio-polymer nanocomposites.

Nanolaminates consist of two or more bonded (physical/chemical) layers of nanomaterial, and are well suited for packaging of fruits and vegetables. They are also helpful in formulating edible coatings, biofilms, and foamings that are current needs of the fruit and vegetable industry (Table 2.5). These coatings or films would act as a barrier to moisture, lipid, or gasses and can increase the textural properties of foods (Figure 2.3); or act as carriers of functional agents including antioxidants, nutrients, and antimicrobials [73]. Currently proteins, polysaccharides, and lipids are being used for fabricating these types of films and coatings.

Composite materials comprise more than one phase, where in one is a continuous phase (polymer) and the other is a disperse phase (filler or composite material). In general, nanocomposites are referred to those comprising of a single or a mixture of polymers with at least one organic or inorganic filler with less than 100 nm dimensions. Several nanocomposites are incorporated into the packaging film (such as: SiO_2, clay, $KMnO_4$, nanocellulose, SiC, nanofibrillated cellulose (NFC), CNTs, etc.), that could enhance their mechanical (stiffness, toughness, tensile strength, shear strength, delamination resistance, fatigue, thermal stability) and barrier properties.

TABLE 2.4　Nanocomposites for Packaging of Horticultural Produce

Composite	Function	References
Clay nanocomposite	Higher tensile properties and lower water vapor transmission rate	[84]
Ethylene vinyl alcohol copolymer containing nanoclays	Increases barrier properties and shelf-life of fruits and Vegetables	[54]
Montmorillonite 3–5%	Light weight, stronger with more thermal stability, and increased barrier properties	[88]
Organic substance: protein, peptide, or lipid	Increased toughness of a packaging material	[88]
PET coated with silica and alumina	Improved moisture and gas barrier, transparent	[71]
Potato starch and calcium carbonate	Improves thermal stability and also biodegradable	[72, 104]
Silicate	Improved gas barrier, mechanical strength, heat resistance	[7, 41, 94]

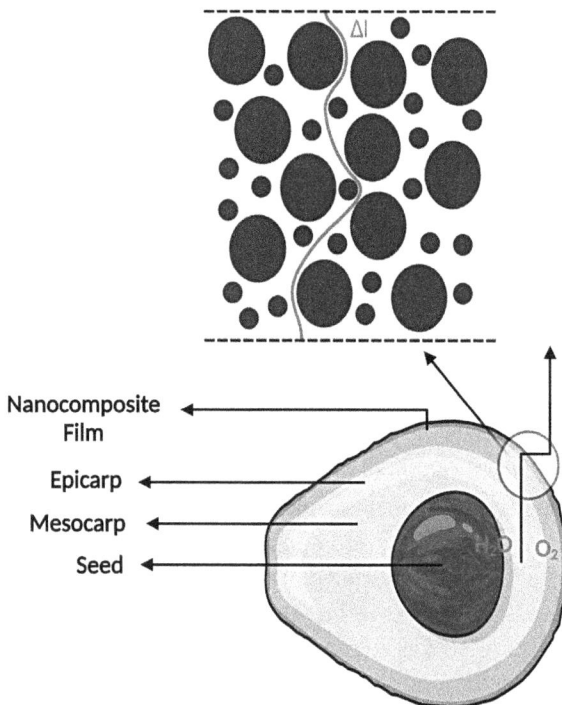

FIGURE 2.3　Tortuous path of water movement from a nanocomposite film or coating in fruits and vegetables.

TABLE 2.5 Application of Various Nanocomposites for Horticultural Crops

Crop	Polymer Composite Used	Function	References
Acerola	Fruit purees and alginate: incorporated into cellulose whiskers (CW) or montmorillonite (MMT) as edible film	• Decreased fruit weight loss, decay incidence, and ripening rates • Ascorbic acid retention	[3]
Apple and lettuce	Ag nanoparticles incorporated into ethylene-vinyl alcohol copolymer (EVOH) films	• Inhibits the growth of L. monocytogenes	[67]
Barberry	AgNPs + LDPE	• Reduced microbial spoilage	[75]
Cherry tomatoes, lychee, and grapes	Carvacrol and alumino silicate incorporated into polyamide (Nylon 6) film	• Broad spectrum antifungal activity against *Alternaria alternata*, *Botrytis cinerea*, *Penicillium digitatum* and *Aspergillus niger*	[101]
Fresh apples, fresh carrot, fresh orange juice	Ag, TiO$_2$ incorporated into polyethylene	• Inhibition of *E. coli*, listeria growth	[69]
Fresh-cut carrots	Calcium alginate + Ag-montmorillonite	• Reduced decay or rot	[19]
Fresh-cut melon	Cellulose + AgNPs	• Reduced total mesophilic aerobic bacteria, psychro-trophic bacteria, yeasts, molds	[30]
Kiwifruit	Nano-Ag, nano-TiO$_2$, and montmorillonite incorporated in polyethylene film	• Decreased oxygen and water vapor permeability • Reduced weight loss, Spore germination	[42]
Mushroom-Flammulina velutipes	Nano-Ag, nano-TiO$_2$, attapulgite, nano-SiO$_2$ incorporated into poly ethylene and LDPE	• Reduced weight loss and nutrient loss • Limits the ROS accumulation and lipid peroxidation	[22]

TABLE 2.5 (*Continued*)

Crop	Polymer Composite Used	Function	References
Orange juice	LDPE films with Ag and ZnO nanoparticles	• Reduced microbial load below the threshold level up to 28 days	[26]
		• Reduced ascorbic acid degradation and brown pigmentation	
Pears, carrots	Sodium alginate + AgNPs	• Reduced spoilage from *E. coli, S. aureus*	[28]
Strawberry	Nano-silver and Nano-silicate composites are incorporated into polyethylene and polypropylene	• Retains high ascorbic acid and phenolic content	[120]
		• Reduces the fruit decay and weight loss	

1. **Preparation of Nanocomposites:** Nanocomposites are fabricated using nanoreinforced polymer layers to enhance barrier, mechanical, and thermal properties (Figure 2.4). However, interaction between reinforced nanomaterials and polymers is the critical factor in formulation of a nanocomposite. The common defects present in the macroscopic reinforcing components shall become negligible when their size is reduced to nanoscale level [64]. Clay/ polymer nanocomposites are most studied nanocomposites for majority of the food packaging applications. Hence, these newly developed biodegradable polymer-based nanocomposites are the next generation materials acting like green nanocomposites and are considered as wave of the future.

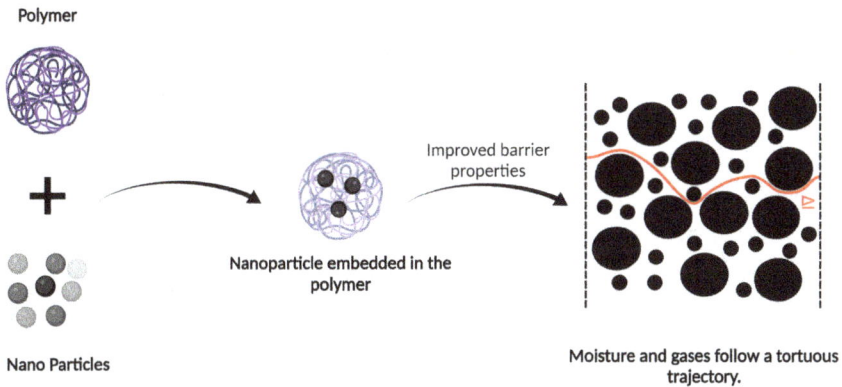

Polymer

+

Nano Particles

Nanoparticle embedded in the polymer

Improved barrier properties

Moisture and gases follow a tortuous trajectory.

FIGURE 2.4 Synthesis of nanocomposites with improved barrier properties.

2. **Functional Benefits of Nanocomposite Packaging (Figure 2.5):**

   i. **Antimicrobial Ability:** The antimicrobial NPs are used to kill the microorganisms, which are present on food or packaging material. Apart from antimicrobial activity, oxygen scavengers and UV scavengers are also used as nano tools for extending the marketable value of the product.

   ii. **Eco-Friendly Packaging Material:** The nanocomposite packaging materials are non-toxic and biodegradable.

iii. **Improved Strength of Package:** The nanocomposite incorporated polymeric matrix package can improve the barrier properties against oxygen, water, temperature, and humidity.

iv. **Indicator Tag:** Intelligent/ Smart packaging systems with attached labels/ incorporated into/ printed onto a food packaging material offer greater possibilities of monitoring product quality, tracing of critical points, and in providing more detailed information throughout the supply chain.

v. **Nano-Sensing:** Food packages equipped with nanosensors are designed for control of internal and external environments. This novel concept is intended for biochemical sensing of microbial changes in the food. For instance, detection of specific pathogens developing in the food, or specific gasses generated from food spoiling.

vi. **Self-Cooling Packaging:** The micro-sized powered systems could make use of a flexible or thin-film photovoltaic cell for cooling processed foods by using thermoelectric materials. This technology shall reduce the need for large-scale and long-time refrigeration in the supply chain, although it may generate a higher cost in this case.

2.3.4.1.2   Nano Biosensors

The chemical sensors translate chemical data into signals contains analytical information. The sensing capacity of chemical sensors depends on specific molecule concentration in the sample. The components of sensors include receptor-recognition system and a transducer. However, nano biosensors are the analytical devices employing a biological material as recognition molecules integrated within a physicochemical transducer or in intimate contact with the transducing microsystems. Biosensors contain optoelectronic recognition system which utilizes cells, enzymes, or organelles mediated biochemical reactions [78].

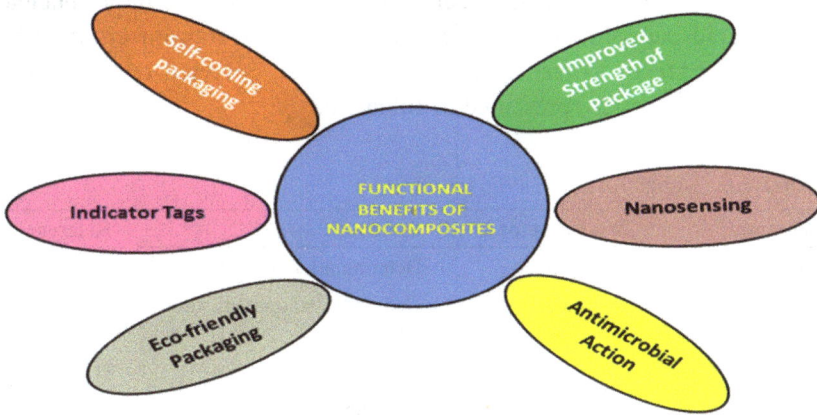

FIGURE 2.5 Functional benefits of nanocomposite packaging.

Nano biosensors equipped packages are also intended for tracking of the internal or/and external conditions of food products, pellets, and containers, throughout the supply chain and these sensors allow the detection of contaminants, spoilage microbes, pathogens, and temperature abuse of the produce during cold chain supply [45]. Ethylene efflux was used for measuring the fruit metabolism in a respiration chamber and also used for detection of ethylene (for apple, avocado, pear, and kiwi) by Blanke et al. [5]. For example, these smart packaging could monitor temperature or humidity over time and provide relevant information on abuse of these conditions, by color change. Nanosensors incorporated into the packaging films can alarm the consumers regarding food spoilage by detecting the gasses given off by the spoiled food or by changing its own color. The use of NPs to develop nanosensors for detection of food contaminants, toxins, and pathogens in food system is getting much popularity now days. Types of nano sensors are described below:

1. **Biosensors:** These are compact, analytical devices that detect, records, and transmit data relating to biochemical reactions (Figure 2.6). They consist of a bio receptor that recognizes a target analyte (which is an organic or biological material such as antigen, microbe, hormone, nucleic acid, and enzyme); and a transducer (which may be of electro chemical, optical, and acoustic form depending on the parameter to be measured) that converts biochemical signals into a quantifiable electrical response [117].

The bio receptors may be either organic (enzyme, hormone, nucleic acid, antigen) or biological (microbe). The transducers may be of optical, acoustic, or electrochemical. Selected commercially biosensors are presented in Table 2.6.

TABLE 2.6　List of Commercial Biosensors

Biosensor	Bio Receptor Used	Function	References
Biogenic amines	Microbe	Determination of amines	[86]
Glucose biosensor	Polyelectrolyte	Detection and quantification of glucose	[66]
Sentinel system®	Antibodies	Detect the food pathogen	[117]
Toxin guard®	Antibodies	Detection of *Salmonella* sp., *Campylobacter* sp., *E. coli*, *Listeria* sp.	[6]
Xanthine	Platinum, silver, and pencil graphite	Detection of xanthine, (adenine nucleotide degradation product in animal tissue)	[2]

Biosensors are widely used to detect, record, and transmit info relating to biological reactions. They are developed by incorporating the NPs into nanostructured transducer of the biosensor devices [110]. Major application of biosensors is for rapid and accurate detection of foodborne pathogens and their toxins; ripeness of the climacteric produce, and temperature abuse suffered by the produce during handling, etc.

The freshness of the horticultural produce can also be detected based on the nucleotide-related compounds using nanosensors to determine whether the food is fit for human consumption. For instance, Toxin Alert (Ontario, California, USA) has developed a diagnostic system called Toxin Guard that incorporates antibodies into plastic packaging film to detect harmful pathogens. Here when the antibody encounters a target pathogen, the packaging material exhibits a clear visual signal to alert the consumer, retailer, or inspector. This can be used to detect gross contamination but not sensitive enough for detecting low levels of disease-causing pathogens [6].

Ripeness sensor marketed as Ripe Sense label senses aromatics or other volatile emissions such as ethylene, from the ripening fruit. The device signals ripeness by label visual cue/color change

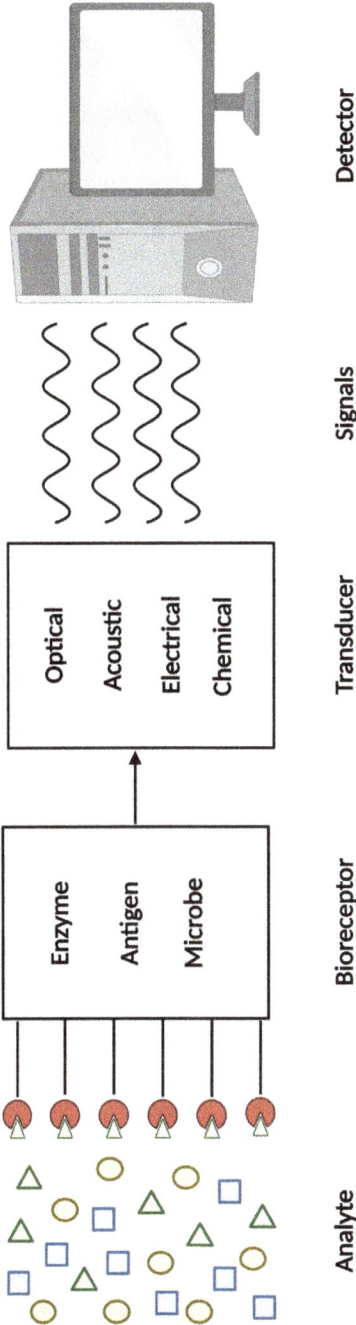

FIGURE 2.6 Schematic representation of working of a biosensor.

for fruit that does not ripe. However, the major drawback of this tool is, instability of the biological sensing component that tends to degrade and lose its effectiveness over a short period of time due to changes in pH, temperature, or ionic strength of its environment.

2. **Electronic Nose (E-Nose):** This is type of aroma-sensor technology mainly designed with electronic chemical gas sensors to detect and distinguish odors in complex samples by mimicking the human nose. This technology was developed by Moncrieff in 1961 to detect the odors. The concept of e-nose as a chemical array sensor system for odor classification was introduced primarily by Persaud and Dodd in 1982 [53]. The electronic nose is also called E-nose/ artificial nose/ sensor array system [114]. The chemical interaction between the volatile compounds and sensor leads to the generation of electrical signals which are recorded in a computer system (Figure 2.7). The recorded signals of sensors are unique for a particular odor compound, which is confirmed further by using chemometric tools and with standards.

The E-nose system comprises 3 major components: (a) sampling system, (b) detection system, and (c) data processing and pattern recognition algorithms. The various sampling systems used are static headspace (SHS) technique, purge, and trap (P&T) technique and dynamic headspace (DHS) techniques; and the commonly used detectors includes conducting polymer (CP) micro-sensors, metal-oxides sensors (MOS), metal oxide semiconductor field effect transistors (MOSFET), bulk acoustic wave (BAW) sensors, optical sensors. Lastly, the perceived electronic signals are processed by using various data analysis techniques including principal component analyzes (PCA), linear discriminant analysis (LDA), hierarchical cluster analysis (HCA) [51].

The main drawback of E-nose system is its sensitivity to environmental conditions like temperature and relative humidity [51, 53]. This E-nose technology has multifarious applications in the field of horticulture (Table 2.7), e.g.: judging the maturity of banana, melon; analysis of quality components of blueberry; determination of maturity in sapodilla using pattern recognition system; estimation of volatile compounds in orange and grapefruit juice; and also, in the volatile profile analysis of the vegetable like tomato. Werlein and Watkinson [112] compared the sensory

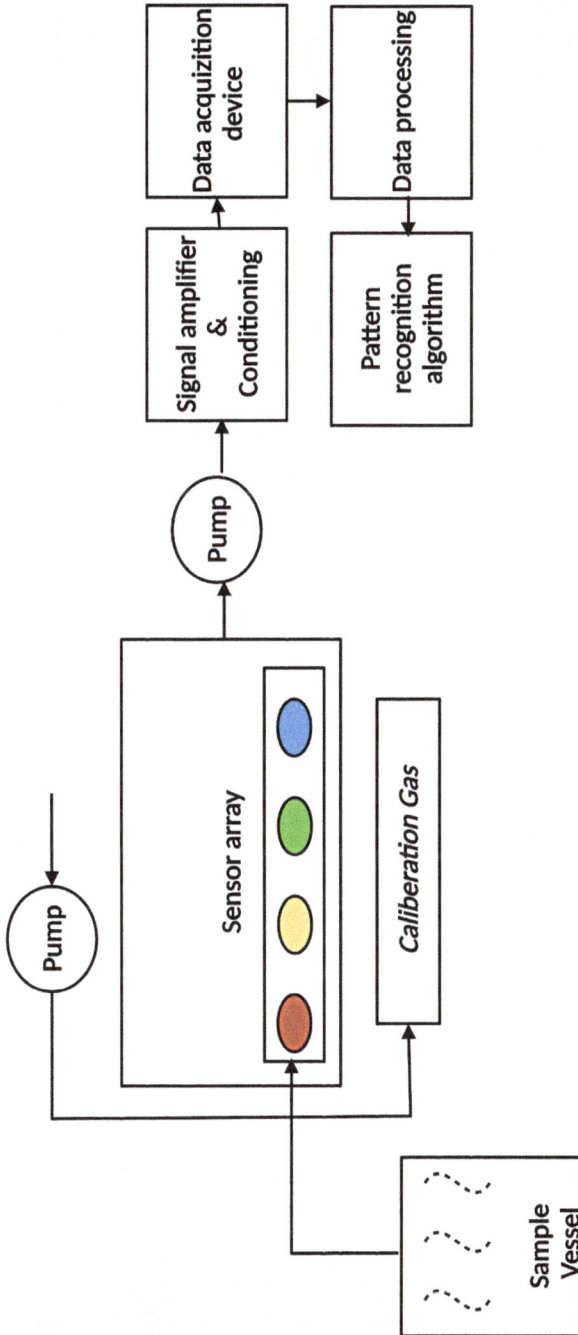

FIGURE 2.7   The components and working of an e-nose system.

quality of conventionally processed carrots, green beans, and potatoes using metal oxide sensors and sensory panels.

TABLE 2.7 Applications of E-Nose in Horticulture Crops

Crop	Type of Sensor	Data Analysis Tool	Objectives
Angelica gigantis	MOS	PCA and ANN	• Discrimination of *Angelica gigantis* radix
Cocoa	MOS	ANN	• Identification of the key aroma compounds in cocoa powder
Coffee	CP and MOS	PCA	• Coffee quality analysis • Evaluating coffee ripening
Ginsengs	MOS	DFA and PCA	• Investigation of changes in aroma of ginsengs of different growing years
Jasmine flowers	MOS	LDA	• Characterization of the volatile aroma compounds from Jasmine flowers grown in India
Olive oil	CP	LDA	• Characterization of olive oil
	MOS	LDA, KNN, ANN, and SIMCA	• Discrimination of olive and seed oils
Saffron	MOS	PCA	• Determination of the volatile profile
Tea	MOS	PCA and ANN	• Classification of tea aroma
	MOS	PCA and LDA	• Monitoring of black tea fermentation process
White Pepper	MOS	LDA	• Sensory testing to assess flavor quality of white pepper
Zingiberaceae	MOS	PCA	• Identify 10 different species of Chinese Herbal Medicines

2.4   ROLE OF NANOTECHNOLOGY IN PROCESSING OF HORTICULTURAL PRODUCE

In horticultural food processing, nanotechnology can be applied in two different approaches [89]:

1. **The Top-Down Method** (Figure 2.8): This involves the process (physical/chemical) of breaking larger food particles into smaller ones of nanometric dimensions [20]. Grinding and milling are

the major mechanisms used to produce such nanomaterials. For example, dry milling of green tea powder is done to enhance its antioxidant activity. This technique reduces the powder size to 1,000 nm leading to high ratio of nutrient digestion and absorption thus promoting the activity of oxygen eliminating enzyme [89].

2. **Bottom-Up Method** (Figure 2.8): This involves manipulation of individual atoms and molecules into nanostructures comprising of discrete functional parts, either inside or on the surface, of which one or more are on the nano-scale [48]. This method could create more complex molecular structures out of biological material using methods such as crystallization, layer-by-layer (L-b-L) deposition, and self-assembly [20]. For example; organization of starch molecules and the folding of globular proteins and protein aggregates are self-assembly structures that form stable entities [89].

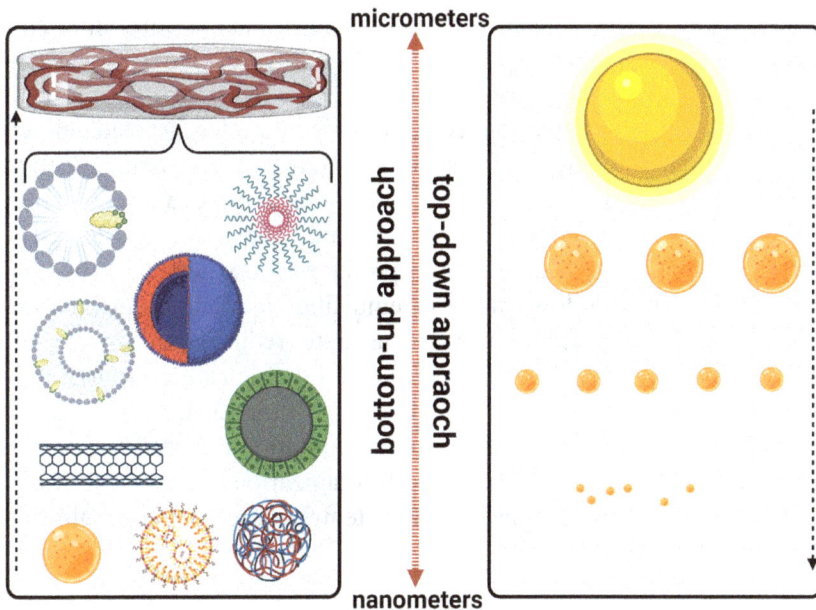

FIGURE 2.8   Two processes of nanoencapsulation: top-down; bottom-up.

2.4.1   NANOENCAPSULATION

Nanoencapsulation is a process, in which micro or nano bioactive materials (core) are enveloped with a continuous film of polymeric material (the shell) to produce capsules in the micrometer or millimeter range. Encapsulation can be achieved using number of chemical or physical techniques, such as: interfacial polymerization, extrusion spherization, and spray drying [63]. Delivery of bioactive compounds in the body system depends mainly on their size. The nanoencapsulation has greater advantage over the microencapsulation due to: enhancement of their bioavailability, improved controlled release, and enabled precision targeting of the bioactive compounds to a greater extent [27].

Recently various bioactive compounds are being encapsulated to reduce their wastage and preventing the disease incidence. Reducing the particle size would enhance the bioavailability, delivery properties, and solubility of the nutraceuticals because of increase in their surface area per unit volume and thus their biological activity [100]. The bioactive compounds are classified into two types lipophilic (beta-carotene, docosahexaenoic acid (DHA), lycopene, lutein, and phytosterols) and hydrophilic (ascorbic acid, polyphenols, etc.). Nano-delivery vehicles such as solid lipid NPs, nanoemulsions, spray dried NPs and modified polymer matrices are mostly used in food industry for encapsulation [14].

Various nanoencapsulation techniques are in use for targeted delivery of the bioactive compounds, such as: coacervation, emulsification, emulsification-solvent evaporation, inclusion complexation, nanoprecipitation, and supercritical fluid technique. These are considered as nanoencapsulation techniques as they produce nanoscale range (10–1,000 nm) capsules. In top-down process, emulsification, and emulsification-solvent evaporation techniques are used to reduce the particle size and structure shaping for development of suitable stable product [27]. Whereas in bottom-up process, supercritical fluid technique, inclusion complexation, coacervation, and nanoprecipitation are used to constrict the materials by self-assembly and self-organization of molecules, which are influenced by many aspects such as temperature, pH, concentration, and ionic strength [27].

2.4.2 NANOFILTRATION

Nanofiltration is a process of concentrating juices driven by pressure, whereas filtration efficiency is affected by sieving and Donnean (charge) effects. This technique appears as an imperative substitute to orthodox means of food processing. It lies in-between the reverse osmosis (RO) and UF (ultra-filtration) process in terms of their separation character-istics. It purifies the organic solutes with molecular weight between 100 and 1,000 Da and uses pressures between 1 and 4 MPa. In this technique, the nanofilters carry negative charge on their surface, resulting in the attraction of the positively charged ions and repulsion of the negatively charged ions, which is popularly known as Donnean effect. Nanofiltra-tion is used for various purposes in juice industry, viz. to recover aromas, wastewater treatment from juice beverage production and to regulate sugar concentration [24].

2.4.3 NANO BAR-CODES

Use of barcode is getting much popularity now a days both in the domestic as well as international market for selling of fruits and vegetables. The use of nano-barcodes on food products can be used to trace outbreaks by following unique tags that are placed on individual products. The addition of nanocarbohydrates is another method that can be used to detect and eliminate contaminants in food products in order to limit the amount of contaminated food products that is packaged and distributed to the public. The bar provides the information about production, place of packaging and price, etc. The nano bar-codes also function similar to the conventional bar codes. The nano-bar codes fluorescent under UV light when target compounds are detected. They improve the package security, assures brand and authenticity [87]. This technology has a number of competitive advantages, viz:

- Difficult to counterfeit;
- Extraordinary durability/compatibility;
- Low cost of manufacture and implementation;
- Unlimited numbers of unique machine-readable code.

2.5 RISK ASSESSMENT AND SAFETY CONCERNS

Nanotechnology also exposes consumers to certain insoluble and feasibly bio-persistent NPs (often termed as hard nanomaterials) through drinks (beverages) and food products. The major concern of these nanomaterials after entering into our body by crossing the biological barriers is to reach the sensitive parts our body, which are otherwise not accessible in their macro forms. There is very little understanding about the potential risks of functionalized nano(bio)-materials. Though most of the food materials processed at nano-scale may not compulsorily raise any health concerns, yet there are number of knowledge gaps in understanding of the properties and behavior of the hard nanomaterials, which might be used in various horticultural applications. There is a potential risk from the nano-technology derived food contact materials, wherein the risk level depends on the migration behavior of the nanomaterials from the packaging material into the food. Such gaps in the current understanding make it difficult for assessing the risk of nanomaterials to the consumers [10].

As with any new technology, there would be certain unanticipated adverse effects along with the envisioned and subsidiary benefits of these applications. Still there is a great deal in learning about the nutritional and safety penalties of introducing nanosized materials into foods and food packaging materials. For instance, how does the properties of nanomaterials alter when they are incorporated into different forms of food matrices or migrate from packaging materials into foods? What would happen when nanomaterials interact with a distinctive biological system such as the human gut? Thus, for developing nanotechnology into a safe, effective tool for use in horticultural science, it requires scientific addressing of all these queries.

2.6   REGULATORY ISSUES

Since the applications of nanotechnology were indispensable, most of the regulatory challenges are not specific to food and other related sectors or to the developing countries alone. Indeed, these challenges require huge efforts at international level for realization of their potential in such a way as to be beneficial and safe to the consumers. For this, establishment of international research collaborations and networks is highly essential to address various aspects of the existing and new applications of nanotechnology in horticulture and related food sectors. Industry sponsored research through PPP (public-private-partnership) modes should be encouraged.

To overcome the present barriers in horticulture, internal collaborations among R&D institutes, industries, and government departments should be encouraged. Clear and consistent guidelines for risk assessment of the nano products should be developed. A global body must be established to ensure quality control and safety assessment of the nano-based products. The nano-based products could be possibly labeled for informing to the customers. Ultimately, we need to develop a well harmonized regulatory system to reach the nutritional security, nanotechnology is a viable option and further studies are vital to examine the threats of nanomaterials on human health [12, 35, 40].

2.7 SUMMARY

Nanotechnology has a great potential to address the complex technical issues relating to perishable food chain supply. Horticultural supply chain management is highly complex because of the varied characteristics of horticultural commodities. The application of nanotechnology is indispensable in the packaging followed by product tracking, tracing, storage, and distribution. Thus, the nanotechnology has great potential for reducing the postharvest losses, improving product quality, and increasing the product compactivity in the domestic and international market. It also promises numerous exciting changes for augmentation of health, wealth, and the quality of life with low level impact on the environment. These promising applications of nanotechnology deals with low use efficiency of farm inputs, abiotic, and biotic stresses to achieve sustainable horticulture systems.

KEYWORDS

- abiotic stresses
- aquaporins
- Cladosporium
- Colletotrichum
- expression profiling
- green synthesis
- isomerization
- nanoenzyme

- paraquat
- pseudomonas
- weedicides
- Xanthomonas

REFERENCES

1. Aminabhavi, T. M., Kulkarni, A. R., Soppimath, K. S., Dave, A. M., & Mehta, M. H., (1999). Applications of sodium alginate beads crosslinked with glutaraldehyde for controlled release of pesticides. *Polymer News, 24*, 285–286.
2. Arvanitoyannis, Ioannis, S., & Alexandros, S., (2012). Application of modified atmosphere packaging and active/smart technologies to red meat and poultry: A review. *Food and Bioprocess Technology, 5*(5), 1423–1446.
3. Azeredo, H. M. C., Miranda, K. W. E., Hálisson, L. R., Morsyleide, F. R., & Nascimento, M. D., (2012). Nanoreinforced alginate-acerola puree coatings on acerola fruits. *Journal of Food Engineering, 113*(4), 505–510.
4. Barik, T. K., Sahu, B., & Swain, V., (2008). Anosilica from medicine to pest control. *Parasitology Research, 103*(2), 253–258.
5. Blanke, M. M., & Shekarriz, R., (2010). Gold nanoparticles and sensor technology for sensitive ethylene detection. In: *XXVIII International Horticultural Congress on Science and Horticulture for People* (*IHC2010, 934): International Symposium* (pp. 255–262).
6. Bodenhamer, W. T., George, J., & Davies, E., (2004). *Surface Binding of an Immunoglobulin to a Flexible Polymer using a Water-Soluble Varnish Matrix* (p. 21). U.S. Patent 6,2004,17,692-973.
7. Brody, A. L., (2006). *Nano and Food Packaging Technologies Converge* (Vol. 60, p. 3). Food Technology Magazine. Online: https://www.ift.org/news-and-publications/food-technology-magazine/issues/2006/march/columns/packaging (accessed on 06 October 2021).
8. Canas, J. E., Monique, L., Shawna, N., Rodica, V., Lenore, D., Mingxiang, L., Ramya, A., et al., (2008). Effects of functionalized and non-functionalized single-walled carbon nanotubes on root elongation of select crop species. *Environmental Toxicology and Chemistry, 27*(9), 1922–1931.
9. Castro, M. J. L., Carlos, O., & Alicia, F. C., (2013). Advances in surfactants for agrochemicals. *Environmental Chemistry Letters, 12*(1), 1–11.
10. Chaudhry, Q., & Laurence, C., (2011). Food applications of nanotechnologies: An overview of opportunities and challenges for developing countries. *Trends in Food Science and Technology, 22*(11), 595–603.
11. Chaudhry, Q., Laurence, C., & Richard, W., (2010). Nanotechnologies in food. *Royal Society of Chemistry, 2010*, 230–240.

12. Chaudhry, Q., Michael, S., James, B., Bryony, R., Alistair, B., Laurence, C., Robert, A., & Richard, W., (2008). Applications and implications of nanotechnologies for the food sector. *Food Additives and Contaminants, 25*(3), 241–258.
13. Chen, H., & Yada, R., (2011). Nanotechnologies in agriculture: New tools for sustainable development. *Trends in Food Science and Technology, 22*(11), 585–594.
14. Chen, L. G. E., & Remondetto, M. S., (2006). Food protein-based materials as nutraceutical delivery systems. *Trends in Food Science and Technology, 17*(5), 272–283.
15. Chin, C. P., Ho-Shing, W., & Shaw, S. W., (2011). New approach to pesticide delivery using nanosuspensions: Research and applications. *Industrial and Engineering Chemistry Research, 50*(12), 7637–7643.
16. Chinnamuthu, C. R., & Kokiladevi, E., (2007). Weed management through nanoherbicide. *Application of Nanotechnology in Agriculture, 10*, 978–981.
17. Chookhongkha, N., Sopondilok, T., & Photchanachai, S., (2012). Effect of chitosan and chitosan nanoparticles on fungal growth and chilli seed quality. *International Conference on Postharvest Pest and Disease Management in Exporting Horticultural Crops-PPDM, 973*, 231–237.
18. Cifuentes, Z., Laura, C., Jesus, M. D., Clara, M., Ricardo, I. M., Diego, R., & Alejandro, P., (2010). Absorption and translocation to the aerial part of magnetic carbon-coated nanoparticles through the root of different crop plants. *Journal of Nanobiotechnology, 8*(1), 26–34.
19. Costa, C., Conte, A., Buonocore, G. G., Lavorgna, M., & Del, N. M. A., (2012). Calcium-alginate coating loaded with silver-montmorillonite nanoparticles to prolong the shelf-life of fresh-cut carrots. *Food Research International, 48*(1), 164–169.
20. Cushen, M., Kerry, J., Morris, M., Cruz-Romero, M., & Cummins, E., (2012). Nanotechnologies in the food industry-recent developments, risks and regulation. *Trends in Food Science and Technology, 24*(1), 30–46.
21. Deepa, M., Palagiri, S., Kandula, V. N., Kota, B. R., Thimmavajjula, G. K., Tollamadugu, N., & Venkata, K. V. P., (2015). First evidence on phloem transport of nanoscale calcium oxide in groundnut using solution culture technique. *Applied Nanoscience, 5*(5), 545–551.
22. Donglu, F., Yang, W., Benard, M. K., An, X., Hu, Q., & Zhao, L., (2016). Effect of nanocomposite packaging on postharvest quality and reactive oxygen species metabolism of mushrooms (*Flammulina velutipes*). *Postharvest Biology and Technology, 119*, 49–57.
23. Doyle, M. E., (2006). *Nanotechnology: Brief Literature Review*. Food Research Institute Briefings. Online http://fri.wisc.edu/docs/pdf/FRIBrief_Nanotech Lit_Rev.pdf (accessed on 05 October 2021).
24. Echevarria, A. P., Torres, C., Pagan, J., & Ibarz, A., (2011). Fruit juice processing and membrane technology application. *Food Engineering Reviews, 3*(3, 4), 136–158.
25. Elek, N., Roy, H., Uri, R., Roy, R., Isaac, I., & Shlomo, M., (2010). Novaluron nanoparticles: Formation and potential use in controlling agricultural insect pests. *Colloids and Surfaces A: Physicochemical and Engineering Aspects, 372*(1), 66–72.
26. Emamifar, A., Mahdi, K., Mohammad, S., & Sabihe, S., (2010). Evaluation of nanocomposite packaging containing Ag and ZnO on shelf life of fresh orange juice. *Innovative Food Science and Emerging Technologies, 11*(4), 742–748.

27. Ezhilarasi, P. N., Karthik, P., Chhanwal, N., & Anandharamakrishnan, C., (2013). Nanoencapsulation techniques for food bioactive components: A review. *Food and Bioprocess Technology, 6*(3), 628–647.

28. Fayaz, A. M., Balaji, K., Girilal, M., Kalaichelvan, P. T., & Venkatesan, R., (2009). Myco-based synthesis of silver nanoparticles and their incorporation into sodium alginate films for vegetable and fruit preservation. *Journal of Agricultural and Food Chemistry, 57*(14), 6246–6252.

29. Feng, B., & Liu-Fen, P., (2012). Synthesis and characterization of carboxymethyl chitosan carrying ricinoleic functions as an emulsifier for azadirachtin. *Carbohydrate Polymers, 88*(2), 576–582.

30. Fernandez, A., Pierre, P., & Elsa, L., (2010). Cellulose-silver nanoparticle hybrid materials to control spoilage-related microflora in absorbent pads located in trays of fresh-cut melon. *International Journal of Food Microbiology, 142*(1), 222–228.

31. Fernandez-Perez, M., Garrido-Herrera, F. J., & Gonzalez-Pradas, E., (2011). Alginate and lignin-based formulations to control pesticides leaching in a calcareous soil. *Journal of Hazardous Materials, 190*(1), 794–801.

32. Flores-Cespedes, F., Figueredo-Flores, C. I., Daza-Fernaandez, Fernando, V., Matilde, V., & Manuel, F., (2012). Preparation and characterization of imidacloprid lignin-polyethylene glycol matrices coated with ethylcellulose. *Journal of Agricultural and Food Chemistry, 60*(4), 1042–1051.

33. Frandsen, M. V., Pedersen, M. S., Zellweger, M., Gouin, S., Roorda, S. D., & Phan, T. Q. C., (2010). *Piperonyl Butoxide and Deltamethrin Containing Insecticidal Polymer Matrix Comprising HDPE and LDPE* (p. 23). Patent number WO2010015256A2. Online https://patentimages.storage.googleapis.com/8e/f5/69/23de7e095ab746/WO 2010015256A2.pdf (accessed on 05 October 2021).

34. Gan, L., Wenyao, X., Maosheng, J., Binghui, H., & Manjing, S., (2010). A study on the inhibitory activities of nano-silver to *Xanthomonas campestris* pv. *campestris*. *Acta Agriculturae Universitatis Jiangxiensis, 32*(3), 493–497.

35. Gergely, A., Diana, B., & Qasim, C., (2010). Small ingredients in a big picture: Regulatory perspectives on nanotechnologies in foods and food contact materials. Chapter 10. In: Chaudhry, Q., Laurence, C., & Richard, W., (eds.), *Nanotechnologies in Food* (Vol. 42, pp. 150–181). London: Royal Society of Chemistry. Nanoscience Series.

36. Giannousi, K., Avramidis, I., & Dendrinou-Samara, C., (2013). Synthesis, characterization and evaluation of copper based nanoparticles as agrochemicals against *Phytophthora infestans*. *RSC Advances, 3*(44), 21743–21752.

37. Goswami, A., Indrani, R., Sutanuka, S., & Nitai, D., (2010). Novel applications of solid and liquid formulations of nanoparticles against insect pests and pathogens. *Thin Solid Films, 519*(3), 1252–1257.

38. Gupta, S., Rajesh, K., & Abdelwahab, O., (2013). Formulation strategies to improve the bioavailability of poorly absorbed drugs with special emphasis on self-emulsifying systems. *ISRN Pharmaceutics, 2013*, 16–21.

39. Hellmann, C., Andreas, G., Joachim, & Wendorff, H., (2011). Design of pheromone releasing nanofibers for plant protection. *Polymers for Advanced Technologies, 22*(4), 407–413.

40. Hodge, G. A., Diana, B., & Karinne, L., (2007). *New Global Frontiers in Regulation: The Age of Nanotechnology* (p. 448). Northampton - MA: Edward Elgar Publishing.

41. Holley, C., (2005). Nanotechnology and packaging: Secure protection for the future. *Verpackungs Rundschau (Packaging Review), 56*, 53–56.

42. Hu, Q., Yong, F., Yanting, Y., Ning, M., & Liyan, Z., (2011). Effect of nanocomposite-based packaging on postharvest quality of ethylene-treated kiwifruit (*Actinidia deliciosa*) during cold storage. *Food Research International, 44*(6), 1589–1596.

43. Isiklan, N., (2004). Controlled release of insecticide carbaryl from crosslinked carboxy methyl cellulose bezads. *Fresenius Environmental Bulletin, 13*(6), 537–544.

44. Jana, T., Bidhan, C. R., & Sukumar, M., (2001). Biodegradable film: 6. Modification of the film for control release of insecticides. *European Polymer Journal, 37*(4), 861–864.

45. Jerish, J. J., & Dhinesh, V. K., (2015). Nanosensors and their applications in food analysis: A review. *The International Journal of Science and Technoledge, 3*(4), 80–88.

46. Jha, Z., Neha, B., Shiv, N. S., Chandel, G., Sharma, D. K., & Pandey, M. P., (2011). Nanotechnology: Prospects of agricultural advancement. *Nano Vision, 1*(2), 88–100.

47. Johnston, C. T., (2010). Probing the nanoscale architecture of clay minerals. *Clay Mineral, 45*(3), 245–279.

48. Joseph, T., & Morrison, M., (2006). *Nanotechnology in Agriculture and Food.* Institute of Nanotechnology - Nanoforum Report. Online; https://nanotech.law.asu.edu/Documents/2009/09/nanotechnology_in_agriculture_and_food_234_2644.pdf (accessed on 14 October 2021).

49. Kah, M., Beulke, S., Tiede, K., & Hofmann, T., (2013). Nanopesticides: State of knowledge, environmental fate, and exposure modeling. *Critical Reviews in Environmental Science and Technology, 43*(16), 1823–1867.

50. Khodakovskaya, M. V., De Silva, K., Alexandru, S. B., Enkeleda, D., & Hector, V., (2012). Carbon nanotubes induce growth enhancement of tobacco cells. *ACS Nano, 6*(3), 2128–2135.

51. Kiani, S., Minaei, S., & Ghasemi-Varnamkhasti, M., (2016). Application of electronic nose systems for assessing quality of medicinal and aromatic plant products: A review. *Journal of Applied Research on Medicinal and Aromatic Plants, 3*(1), 1–9.

52. Kok, F. N., Wilkins, R. M., Cain, R. B., Arica, M. Y., Alaeddinoglu, G., & Hasirci, V., (1999). Controlled release of aldicarb from lignin loaded ionotropic hydrogel microspheres. *Journal of Microencapsulation, 16*(5), 613–623.

53. Korel, F., & Balaban, M. O., (2003). Uses of electronic nose in the food industry. *Gıda Teknolojisi Dergisi (Food Technology Magazine), 28*(5), 505–511.

54. Lagaron, J. M., Cabedo, L., Cava, D., Feijoo, J. L., Gavara, R., & Gimenez, E., (2005). Improving packaged food quality and safety, part 2: Nanocomposites. *Food Additives and Contaminants, 22*(10), 994–998.

55. Lao, S. B., Zhi-Xiang, Z., Han-Hong, X., & Gang-Biao, J., (2010). Novel amphiphilic chitosan derivatives: Synthesis, characterization and micellar solubilization of rotenone. *Carbohydrate Polymers, 82*(4), 1136–1142.

56. Lasic, D. D., (1993). *Liposomes: From Physics to Applications* (p. 567). Amsterdam: Elsevier.

57. Lauterwasser, C., (2005). *Small Sizes That Matter: Opportunities and Risks of Nanotechnologies.* Report in Cooperation with the OECD International Futures Program. http://www.oecd.org/dataoecd/32/1/44108334.pdf (accessed on 05 October 2021).

58. Li, H., Feng, L., Lin, W., Jianchun, S., Zhihong, X., Liyan, Z., Hongmei, X., et al., (2009). Effect of nano-packing on preservation quality of Chinese jujube (*Ziziphus jujuba* Mill. var. *inermis* (Bunge) Rehd). *Food Chemistry, 114*(2), 547–552.

59. Li, M., Qiliang, H., & Yan, W., (2011). A novel chitosan-poly (lactide) copolymer and its submicron particles as imidacloprid carriers. *Pest Management Science, 67*(7), 831–836.

60. Liu, J., Feng-Hua, W., Ling-Ling, W., Su-Yao, X., Chun-Yi, T., Dong-Ying, T., & Xuan-Ming, L., (2008). Preparation of fluorescence starch-nanoparticle and its application as plant transgenic vehicle. *Journal of Central South University of Technology, 15*, 768–773.

61. Liu, Q., Bo, C., Qinli, W., Xiaoli, S., Zeyu, X., Jinxin, L., & Xiaohong, F., (2009). Carbon nanotubes as molecular transporters for walled plant cells. *Nano Letters, 9*(3), 1007–1010.

62. Liu, Y., Laks, P., & Heiden, P., (2002). Controlled release of biocides in solid wood. II. Efficacy against *Trametes versicolor* and *Gloeophyllum trabeum* wood decay fungi. *Journal of Applied Polymer Science, 86*(3), 608–614.

63. Lopez-Rubio, A., Rafael, G., & Lagaron, J. M., (2006). Bioactive packaging: Turning foods into healthier foods through biomaterials. *Trends in Food Science and Technology, 17*(10), 567–575.

64. Luduena, L. N., Alvarez, V. A., & Analía, V., (2007). Processing and microstructure of PCL/Clay nanocomposites. *Materials Science and Engineering, 460*, 121–129.

65. Magnuson, B. A., Jonaitis, T. S., & Jeffrey, W. C., (2011). A brief review of the occurrence, use, and safety of food-related nanomaterials. *Journal of Food Science, 76*(6), 126–133.

66. Manesh, K. M., Hyun, T. K., Santhosh, P., & Gopalan, A., (2008). A novel glucose biosensor based on immobilization of glucose oxidase into multiwall carbon nanotubes-polyelectrolyte-loaded electrospun nanofibrous membrane. *Biosensors and Bioelectronics, 23*(6), 771–779.

67. Martinez-Abad, A., Lagaron, J. M., & Maria, J. O., (2012). Development and characterization of silver-based antimicrobial ethylene-vinyl alcohol copolymer (EVOH) films for food-packaging applications. *Journal of Agricultural and Food Chemistry, 60*(21), 5350–5359.

68. Martin-Ortigosa, S., Valenstein, J. S., Lin, V. S. Y., Trewyn, B. G., & Kan, W., (2012). Gold functionalized mesoporous silica nanoparticle mediated protein and DNA codelivery to plant cells via the biolistic method. *Advanced Functional Materials, 22*(17), 3576–3582.

69. Metak, A. M., Farhad, N., & Stephen, N. C., (2015). Migration of engineered nanoparticles from packaging into food products. *LWT-Food Science and Technology, 64*(2), 781–787.

70. Mondal, K. K., & Mani, C., (2012). Investigation of the antibacterial properties of nanocopper against *Xanthomonas axonopodis* pv. *punicae*, the incitant of pomegranate bacterial blight. *Annals of Microbiology, 62*(2), 889–893.

71. Moore, S., (1999). Packaging-nanocomposite achieves exceptional barrier in films. *Modern Plastics, 76*, 31, 32.

72. Moraru, C. I., Panchapakesan, C. P., Qingrong, H., Paul, T., Liu, S., & Jozef, L. K., (2003). Nanotechnology: A new frontier in food science. *Food Technology, 57*(12), 24–29.

73. Morillon, V., Frederic, D., Genevieve, B., Martine, C., & Andree, V., (2002). Factors affecting the moisture permeability of lipid-based edible films: A review. *Critical Reviews in Food Science and Nutrition, 42*(1), 67–89.

74. Morris, V. J., (2011). Emerging roles of engineered nanomaterials in the food industry. *Trends in Biotechnology, 29*(10), 509–516.

75. Motlagh, N. V., Hamed, M. M. T., & Mortazavi, S. A., (2013). Effect of polyethylene packaging modified with silver particles on the microbial, sensory and appearance of dried barberry. *Packaging Technology and Science, 26*(1), 39–49.

76. Mousavi, S. R., & Rezaei, M., (2011). Nanotechnology in agriculture and food production. *Journal of Applied Environmental Biological Science, 1*(10), 414–419.

77. Nair, R., Saino, H. V., Nair, B. G., Maekawa, T., Yoshida, Y., & Kumar, D. S., (2010). Nanoparticulate material delivery to plants. *Plant Science, 179*(3), 154–163.

78. Nic, M., Jirat, J., & Kosata, B., (2019). *IUPAC Compendium of Chemical Terminology: The Gold Book* (p. 232). Research triangle park: International union of pure and applied chemistry (IUPAC).

79. Oberdorster, G., Oberdorster, E., & Oberdorster, J., (2005). Nanotoxicology: An emerging discipline evolving from studies of ultrafine particles. *Environmental Health Perspectives, 113*(7), 823–839.

80. Ocsoy, I., Mathews, L., Paret, M. A. O., Sanju, K., Tao, C., Mingxu, Y., & Weihong, T., (2013). Nanotechnology in plant disease management: DNA-directed silver nanoparticles on graphene oxide as an antibacterial against *Xanthomonas perforans*. *ACS Nano, 7*(10), 8972–8980.

81. Opara, L., (2004). Emerging technological innovation triad for smart agriculture in the 21st century: Part I. Prospects and impacts of nanotechnology in agriculture. *CIGR E-Journal.*

82. Paret, M. L., Palmateer, A. J., & Knox, G. W., (2013). Evaluation of a light-activated nanoparticle formulation of titanium dioxide with zinc for management of bacterial leaf spot on rosa 'noare'. *Horticulture Science, 48*(2), 189–192.

83. Paret, M. L., Vallad, G. E., Averett, D. R., Jones, J. B., & Olson, S. M., (2012). Photocatalysis: Effect of light-activated nanoscale formulations of TiO$_2$ on *Xanthomonas perforans* and control of bacterial spot of tomato. *Phytopathology, 103*(3), 228–236.

84. Park, H. M., Won-Ki, L., Chan-Young, P., Won-Jei, C., & Chang-Sik, H., (2003). Environmentally friendly polymer hybrids part I mechanical, thermal, and barrier properties of thermoplastic starch/clay nanocomposites. *Journal of Materials Science, 38*(5), 909–915.

85. Pimentel, D., (2009). Pesticides and pest control. In: Peshin, R., & Dhawan, A., (eds.), *Integrated Pest Management: Innovation-Development Process* (p. 690). Netherlands: Springer.

86. Pospiskova, K., Ivo, S., Marek, S., & Gabriela, K., (2013). Magnetic particles-based biosensor for biogenic amines using an optical oxygen sensor as a transducer. *Microchimica Acta, 180*(3, 4), 311–318.

87. Prasad, R., Kumar, V., & Prasad, K. S., (2014). Nanotechnology in sustainable agriculture: Present concerns and future aspects. *African Journal of Biotechnology, 13*(6), 705–713.

88. Rashidi, L., & Kianoush, K. D., (2011). The applications of nanotechnology in food industry. *Critical Reviews in Food Science and Nutrition, 51*(8), 723–730.

89. Ravichandran, R., (2010). Nanotechnology applications in food and food processing: Innovative green approaches, opportunities and uncertainties for global market. *International Journal of Green Nanotechnology: Physics and Chemistry, 1*(2), 72–96.

90. Reddy, P. K., Mamatha, N. C., Naik, P., Srilatha, V., & Kumar, P. P., (2016). Applications of nanotechnology in agricultural sciences. *Andhra Pradesh Journal of Agricultural Science, 2*(1), 1–9.

91. Sohadi, R. S., Mehrnaz, A. A., Samim, M., & Abdin, M. Z., (2013). Plant bio-transformable HMG-CoA reductase gene loaded calcium phosphate nanoparticle: *In vitro* characterization and stability study. *Current Drug Discovery Technologies, 10*(1), 25–34.

92. Sastry, K. R., Rashmi, H. B., Rao, N. H., & Ilyas, S. M., (2010). integrating nanotechnology into agri-food systems research in India: A conceptual framework. *Technological Forecasting and Social Change, 77*(4), 639–648.

93. Scariot, V., Roberta, P., Hilary, R., & De Pascale, S., (2014). Ethylene control in cut flowers: Classical and innovative approaches. *Postharvest Biology and Technology, 97,* 83–92.

94. Schaefer, M., (2005). Double tightness. *Lebensmittel Technic, 37,* 52–55.

95. Scott, N., & Chen, H., (2012). Nanoscale science and engineering for agriculture and food systems. *Industrial Biotechnology, 8*(6), 340–343.

96. Scrinis, G., & Lyons, K., (2007). The emerging nano-corporate paradigm: Nanotechnology and the transformation of nature, food and agri-food systems. *International Journal of Sociology of Agriculture and Food, 15*(2), 22–44.

97. Seo, Y. C., Jeong-Sub, C., Hae-Yoon, J., Tae-Bin, Y., Kyoung-Sook, C., Tae-Woo, L., & Myoung-Hoon, J., (2011). Enhancement of antifungal activity of anthracnose in pepper by nanopaticles of thiamine di-lauryl sulfate. *Korean Journal of Medicinal Crop Science, 19*(3), 198–204.

98. Shakil, N. A., Singh, M. K., Pandey, A., Kumar, J., Parmar, P., Virinder, S., Singh, M. K., et al., (2010). Development of poly (Ethylene Glycol) based amphiphilic copolymers for controlled release delivery of carbofuran. *Journal of Macromolecular Science, Part A: Pure and Applied Chemistry, 47*(3), 241–247.

99. Sharma, S., Ahmad, N., Prakash, A., Singh, V. N., Ghosh, A. K., & Mehta, B. R., (2010). Synthesis of crystalline Ag nanoparticles (AgNPs) from microorganisms. *Materials Sciences and Applications, 1,* 1–7.

100. Shegokar, R., & Muller, R. H., (2010). Nanocrystals: Industrially feasible multifunctional formulation technology for poorly soluble actives. *International Journal of Pharmaceutics, 399*(1), 129–139.

101. Shemesh, R., Maksym, K., Nadav, N., Anita, V., & Ester, S., (2016). Active packaging containing encapsulated carvacrol for control of postharvest decay. *Postharvest Biology and Technology, 118,* 175–182.

102. Silva, D. S., Mariana, C. D. S., Grillo, R., Silva De, M. N. F., Tonello, P. S., De Oliveira, L. C., Cassimiro, D. L., et al., (2011). Paraquat-loaded alginate/chitosan nanoparticles: Preparation, characterization and soil sorption studies. *Journal of Hazardous Materials, 190*(1), 366–374.

103. Sorrentino, A., Gorrasi, G., & Vittoria, V., (2007). Potential perspectives of bio-nanocomposites for food packaging applications. *Trends in Food Science and Technology, 18*(2), 84–95.

104. Siegal, R. D., (1997). Trends in nanoparticles. In: Siegel, R. W., Hwu, E., & Roco, M. C., (eds.), *Nanostructured Materials and Nanodevices in the United States* (pp. 1–14). Baltimore - USA: Int. Technology Res. Inst.
105. Subramanian, K. S., & Tarafdar, J. C., (2011). Prospects of nanotechnology in Indian farming. *Indian Journal Agricultural Sciences, 81*, 887–893.
106. Tarafdar, J. C., Xiang, Y., Wei-Ning, W., Dong, Q., & Pratim, B., (2012). Standardization of size, shape and concentration of nanoparticle for plant application. *Applied Biological Research, 14*, 138–144.
107. Tharanathan, R. N., (2003). Biodegradable films and composite coatings: Past, present and future. *Trends in Food Science and Technology, 14*(3), 71–78.
108. Torney, F., Trewyn, B. G., Victor, S. Y. L., & Kan, W., (2007). mesoporous silica nanoparticles deliver DNA and chemicals into plants. *Nature Nanotechnology, 2*(5), 295–300.
109. Tournas, V. H., (2005). Spoilage of vegetable crops by bacteria and fungi and related health hazards. *Critical Reviews in Microbiology, 31*(1), 33–44.
110. Vo-Dinh, T., Brian, M., & Cullum, D. L. S., (2001). Nanosensors and biochips: Frontiers in biomolecular diagnostics. *Sensors and Actuators B: Chemical, 74*(1), 2–11.
111. Wang, L., Xuefeng, L., Gaoyong, Z., Jinfeng, D., & Julian, E., (2007). Oil-in-water nanoemulsions for pesticide formulations. *Journal of Colloid and Interface Science, 314*(1), 230–235.
112. Werlein, H. D., & Watkinson, B. M., (1997). Evaluation of the sensory quality of processed foods by means of the electronic nose. *Seminars in Food Analysis, 2*, 215–220.
113. West, J. L., & Naomi, J. H., (2000). Applications of nanotechnology to biotechnology: Commentary. *Current Opinion in Biotechnology, 11*(2), 215–217.
114. Wilson, A. D., (2013). Diverse applications of electronic-nose technologies in agriculture and forestry. *Sensors, 13*(2), 2295–2348.
115. Wright, J. E., (1997). *Formulation for Insect Sex Pheromone Dispersion* (p. 18). Patent US: 5670145 A0923.
116. Yadollahi, A., Arzani, K., & Khoshghalb, H., (2009). the role of nanotechnology in horticultural crops postharvest management. In: *Southeast Asia Symposium on Quality and Safety of Fresh and Fresh-Cut Produce* (Vol. 875, pp. 49–56).
117. Yam, K. L., Takhistov, P. T., & Joseph, M., (2005). Intelligent packaging: Concepts and applications. *Journal of Food Science, 70*(1), 1–10.
118. Yang, F. L., Xue-Gang, L., Fen, Z., & Chao-Liang, L., (2009). Structural characterization of nanoparticles loaded with garlic essential oil and their insecticidal activity against *Tribolium castaneum* (Herbst). *Journal of Agricultural and Food Chemistry, 57*(21), 10156–10162.
119. Yu-qin, F. U., Lu-Hua, L. I., Pi-Wu, W. A. N. G., Jing, Q. U., Yong-Ping, F. U., Hui, W. A. N. G., & Jing-Ran, S. U. N., (2012). Delivering DNA into plant cell by gene carriers of ZnS nanoparticles. *Chemical Research in Chinese Universities, 28*(4), 672–676.
120. Zandi, K., Weria, W., Hasan, A., Irajh, B., & Lotfali, N., (2013). Effect of nanocomposite-based packaging on postharvest quality of strawberry during storage. *Bull. Environmental Pharmacology Life Science, 2*(5), 28–36.
121. Zheng, L., Fashui, H., Shipeng, L., & Chao, L., (2005). Effect of nano-TiO_2 on strength of naturally aged seeds and growth of spinach. *Biological Trace Element Research, 104*(1), 83–91.

CHAPTER 3

ROLE OF NANOMATERIALS IN PLANT GROWTH AND NUTRITION

BRIJESH PATIL MUDER PAKEERAPPA, HARSHITA SINGH, and HARSHVARDHAN GOWDA VENKATACHALA

ABSTRACT

The unique characteristics of nanomaterials, such as, surface area to volume ratio, hydrophilicity, and surface charge will be transformed through interaction with abiotic (water, light, radiation, temperature, acidity, minerals) and biotic (microorganisms) soil components. To enhance micronutrients bioavailability to plants, nano-encapsulated micronutrients formulation or nano-minerals supplementation offers a better strategical platform. However, differences in nutrient requirement and interplay between nutrients concentration often limits adequate nutrients for plant growth. Through this chapter, nanomaterials application for plant growth and nutrition is discussed in brief.

3.1 INTRODUCTION

Nanomaterials possess potential application in agriculture, food, biotechnology, manufacturing, and health sectors. The potential application in agriculture includes development of nano-fertilizers for enhanced productivity, nano-based pesticide formulations for plant protection and disease management, nano-enabled sensors for plant and soil health monitoring [43]. In addition, plant nutrients at nanoscale can be absorbed by roots and enhance the efficiency of agricultural practices. The collective approach of recent studies focusing on enhanced efficiency and efficacy of agricultural practices can open doors to sustainable agriculture than conventional practices.

Current approaches of integrating plant growth and nutrition include genetic breeding, stress resistant plant variety development and integrated pest management. The development of successful strategy requires plant nutritional status data at different growth stages. Robust nutrient supplementation often facilitates response of plant growth and higher production. However, differences in nutrient requirement and interplay between nutrients concentration often limits adequate nutrients for plant growth [11, 41].

Consistently, many investigations in the literature have discussed influence of nanomaterials on plant growth, but the mechanism of action is unclear. In addition, the enhanced growth, crop yield, crop germination with nanoparticle exposures has been discussed in literature. However, many studies have compared effect of nanomaterials with their bulk compound counterparts. The positive effect of nanomaterials exposure may depend on the particles size among nano and bulk materials.

Presence of nanomaterials in any environment or matrix highly influences their physical and chemical characteristics and understanding their biological fate; and translocation in soil matrices is highly challenging and possess complexity in soil-plant interaction platforms. The unique characteristics of nanomaterials, such as, surface area to volume ratio, hydrophilicity, and surface charge will transformed through interaction with abiotic (water, light, radiation, temperature, acidity, minerals) and biotic (microorganisms) soil components and eventually they decide the nanomaterials stability, transportation, aggregation, and their availability for plants and other biotic components of soil [34].

Micronutrients are crucial in plant growth and they are actively involved in defense mechanism during disease infestation through production of secondary metabolites. However, element availability in soil is often restricted between neutral to alkaline pH. For instance, iron, manganese, zinc bioavailability decreases with increase in pH [33]. In addition, providing micronutrients to shoot tissues is ineffective due to shoot to root translocation, when micronutrients are not basipetally transferred. Importantly, non-essential inorganic nutrients are essential in crop defense mechanism against diseases. For instance, calcium chloride and orthophosphate application have resulted in reducing pathogen *Fusarium* wilt in *Lycopersicon esculentum*. Unfortunately, metal oxides response pathway is unknown, but identical low bioavailability in soil and root translocation or shoot absorption factors confound efficiency in plant development [5]. Hence to enhance micronutrients bioavailability for plants, nano-encapsulated micronutrients formulation or nano-minerals supplementation offers a better strategical platform.

Engineered nanomaterials are often limited to natural abundant substances, such as, carbon, and silicon. Diverse nanostructures (such as: cylindrical carbon nanotubes (CNTs), multi-layered graphene) have been utilized as biochemical carriers, sensors; and they have helped in understanding various interactive parameters at the plant and soil interface under various biotic and abiotic stresses. In addition, hybrid matrices (such as: protein-protein, protein-polysaccharides) and other combination of materials with organic compounds can be possible alternative matrices for development of nanomaterials.

Broad range of nanomaterials have been utilized as micro- and macronutrients to enhance soil fertility, plant productivity and nutrient uptake rate. Moreover, to achieve increasing food grains and bioenergy demand, we could consider nanotechnology-assisted revolution in agricultural sector. Although it has been reported about the toxicity of nanomaterials on plants, water bodies and soil [44], yet ultimate goal is to reduce hazardous effects of pesticides, fertilizers being utilized in conventional farming.

Conventional agriculture has utilized chemical fertilizers and pesticides, though their great contribution towards food security is being considered, but their side-effects on environment (such as: nutrient imbalance, soil structure destruction, low soil fertility and water body pollution) due to nitrate leaching are demanding for new environmental approach to revive soil ecosystem for plants growth. In addition, release of nutrient in defined rate is essential for targeted delivery. Various nanomaterials have been developed for either nutrient carrying matrix or nutrient itself. For instance, graphene has been developed to enhance the ability of matrix to release the plant nutrient in controlled and steady rate, thereby decreasing higher nutrient requirement and developing sustainable agricultural materials [18].

Nanomaterials possess unique characteristics such as higher surface area to volume ratio, which can help in greater sorption and controlled release of minerals for plant development. In addition, nanomaterials as nano-carriers for many nutrients protects encapsulated material during environmental extremities. Moreover, many reports claim about increasing chlorophyll concentration, thereby enhanced photosynthetic process, amplified secretion of plant development hormones such as gibberellins and decrease in synthesis of abscisic acid with nano-fertilizers application. It has been perceived that, controlled delivery of nutrients at targeted sites provides both nutrients and enhances plant metabolism. Various nano-materials structures such as core-shell geometry, liquid-liquid emulsions, solid-lipid nanoparticles (NPs), liposomes, micelles, and nanotubes have favored controlled release of plant nutrients [1, 35].

The inherent ability of nanomaterial for enhanced surface area to volume ratio enhances the light reaction phase of photosynthesis. Few investigations demonstrate the ability of nanomaterials to absorb and respond to particular wavelength of light, thereby highlighting their optical potential. For instance, nano-titanium dioxide treated spinach accumulated 45% extra chlorophyll-A, which resulted in enhanced photosynthesis and 73% extra dry weight [48].

Nano-titanium dioxide increases uptake of water, oxygen, inorganic nutrients in the cell besides quenching ROS and breakdown of organic compounds. The most significant effect of titanium dioxide nanomate-rials is that it increases light absorbance and activate Rubisco activase, which maximizes the total photosynthetic yield. This ability can translate to higher production of proton carriers, which can effectively carry out improved sugar production phase in the Calvin cycle or dark phase [36]. This intervention can effectively not only make the green leaves photo-synthesize better or faster, but also produce faster or improved food for storage, thereby improving the sink mechanism of prepared sugar.

This chapter briefs on various approaches of nanotechnology being utilized in scientific community for integration of plant development and nutrition. These include and not limited to, as stress management, disease management, fertilizers, metabolic pathway modification, pesticides, and molecular tools [27].

3.2   NANOFERTILIZERS AND NUTRIENTS

Nanofertilizers have been proven their efficient action towards plant nutrition through enhanced nutrient utilization by plant tissues and decreased nutrient loss. Due to their higher surface area and reduced size, absorption by plant tissues is high and plant nutrient can be encapsulated inside nanomaterials, such as: porous substances, nanotubes or coated with biodegradable polymers or NPs itself or nanoemulsions. Moreover, higher surface area to volume ratio helps the nanofertilizer to enter into cell directly, thereby reducing energy intensive uptake pathway compared to conventional fertilizers. Controlled release mechanism involves steady; and stress induced release of fertilizer enhances fertilizer use efficiency and soil fertility by hindering toxic effects associated with over dose of conventional fertilizers.

In addition, higher solubility enhances nutrient solubilization in the rhizosphere and reduces the nutrient loss due to emission as gaseous ammonia or nitrogen oxide and leaching in the form of hydrophilic nitrates. Spherical chitosan NPs have been utilized to encapsulate nitrogen, phosphorus, and potassium from different matrices, such as, urea, calcium phosphate and potassium chloride. The results have demonstrated that nanofertilizer was stable for urea and potassium chloride compared to calcium phosphate. However, lower stability might be result of magnitude of anion in calcium phosphate compared to other nutrients; and presence of positive charge with chitosan different concentration of calcium phosphate can result in formation of stable chitosan nanofertilizer [8].

Preferably, nanoscience may help in devising a mechanism to synchronize with nutrient release and its uptake by plant with controlled or stress induced release strategy. Thus, Nanofertilizers could release nutrients depending on the plant demand while controlling these from premature conversion into other chemical forms, such as: oxides and gas. In addition, Nanofertilizers can also protect the nutrient interacting with water, microorganism, and soil and releasing nutrient only during the plant requirement. Examples for these nano-based strategies are emerging in decent decades.

Organic-inorganic, i.e., alpha-naphthalene acetate with zinc-aluminum hybrid nano-matrix has been utilized by self-assembly strategy. The plant growth regulator release was interpreted based on nitrate or hydroxyl anions concentration in the aqueous solution or alpha-naphthalene acetate anion intercalation in *lamella* host. In addition, there was initial first order kinetics of growth regulator release and controlling through pH modulation [3].

Owing to complexities of controlled release mechanism in soil matrix, urease enzyme has been utilized to modulate nitrogen release by inserting enzyme in nanopores of silica [14]. It is hypothesized that this strategy can be utilized to overcome urease inhibition in the plant system to control ammonia release from urea fertilizer. The industrial preparation of mesoporous materials as fertilizer carriers may be possible, because various types of natural silicates are abundant in nature and have structural similarities to those of synthetic mesoporous materials. Although these strategies are promising, yet they are deficient in mechanism to sense nitrogen level in soil and to respond to plant requirement. In addition to increasing fertilizer utilization efficiency, nanomaterials might be able to enhance the performance, such as: nano anatase-titanium dioxide was used as bactericidal additive to treat spinach, while it could enhance crop dry weight, protein content, chlorophyll, and total nitrogen through photo-reduction of nitrogen gas [45].

Different nanostructures or nutrients itself have been converted to nanoscale for agricultural use. Interestingly, shell-core based nanomaterials offer great tunability for fertilizer. A nanosized manganese carbonate loaded with zinc sulfate was able to regulate the controlled release of zinc; and the hollow shell-core structure served as plant growth medium. The release of zinc from core matrix through dissolution and ion exchange reaction was able to increase rice crop yield and to reduce nutrient loss in soil [46].

Synthetic nanomaterials interaction with biotic and abiotic components in soil matrix are still under investigation. However, present outcomes suggest potential use of nanoscale zinc, iron, copper oxides and manganese as nutrient sources in deficient soils. Increased dry weight and nutrient accumulation has been witnessed with zinc oxide (ZnO) usage in cucumber, peanuts, sweet basil, cabbage, tomato, cauliflower, and chick pea [12, 29, 30, 38, 47].

The release and dissolution rates of water-soluble fertilizers depend on the coating materials. This brings out the idea of developing the entrapped fertilizer within a nanomaterial. Consequently, the fertilizer is protected by the nanomaterial for better survival in inoculated soils, allowing their controlled release into the soil. Research studies have demonstrated efficiency of nanotechnology for increased crop productivity compared to conventional fertilizers with respect to their concentration. Furthermore, climatic change, soil degradation and limited agricultural area for food crop production due to urbanization, access to sufficient food crops from sustainable system is detained [7, 17, 19, 38, 46].

The nanotechnology application for food processing sector is enormous and has been utilized to formulate many functional foods [4, 10, 21, 22, 24, 32, 37].

Nanoparticle and plant interaction can establish system, which enables nutrient controlled delivery, reduced nutrient loss to the environment and reduced phytotoxicity effects. The potential bioaccumulation of nanostructures and transferring to tropical levels is the major concern associated with agricultural nanotechnology. In recent study with cerium oxide, the accumulation was observed in root tissues of soybean and followed by translocation to edible parts of plants [13]. Notably, NPs are also essential plant micronutrients, thereby raising the plant growth through nutritional supplementation.

Moreover, conventional fertilizers formulations often use active nutrients with low-water solubility; and hence, nutrient absorption can be quite low. To overcome low nutrient availability, conventional fertilizers have been used in larger volumes; and leaching, volatilization, and precipitation by soil components cause environmental issues. The nanomaterials have been demonstrated to influence enzyme activities in plant metabolic pathways and electron transport system, thereby enhancing the physiological parameters of plants [23]. Additionally, controlled release property of nanomaterials can assist in synchronizing nutrient flux uptake and release during plant development. This strategy helps in both enhanced nutrient availability and reduces wasteful interactions with air or soil.

3.3 NANOMATERIALS AND SOIL ENVIRONMENT

There are various pathways associated with entry of nanomaterials into soil, which characterize migration and accumulation over time. Nanomaterials can enter into soil with sedimentation in the form of aerosols, dust, direct soil absorption of gaseous components, atmospheric precipitation, and abscission of leaves. After nanomaterials have traveled to water bodies through sewage or industrial emissions, then nanomaterials can accumulate in plant tissues and these are the primary link of a food chain. In a land ecosystem, nanomaterials can accumulate in surface water, ground water, soil, and sewage.

Nanomaterials, such as, titanium dioxide, ZnO, fullerene, iron oxides and silver have been used as cosmetic and personal hygiene products, which later can accumulate in surface water and sewage. Catalytic agents, lubricants, and combustion catalysts (such as: cerium oxide, molybdenum), and platinum-based nanomaterials can accumulate in surface water, sewage, and air ecosystems [15]. Nanomaterials used in agrochemical preparations, such as, nanocarriers, Nanofertilizers can accumulate in soil, surface water, ground water and in sewage. NPs utilized in pharmaceutical preparations can accumulate in soil and waste ecosystem, which can later be translocated to soil matrix.

Though, various pathways and sources can define characteristics of nanomaterials, pollutants ultimately enter to soil surface and depending on soil properties, yet nanomaterials fate will be decided. Pollutants stay in soil matrix for longer time period than other objects and their depositing anthropogenic nature follows as secondary pollution in water bodies and environment. The pollution of soil matrix with nanomaterials poses a risk of translocating these to higher food chains, such as: plants, animals, and humans. The entry of any nanomaterials into any biocenosis component can result in the introduction into other components in hierarchy through food chain. Currently, there are not enough research studies on the migration of nanomaterials in soil matrix. However, nanomaterials absorbed or as composition of colloidal structures are known to be transported, such as, colloidal metal salts. The serious risk associated with metal nanomaterials is their ability to migrate in aqueous solution due to their high solubility [9, 40].

The depth of nanomaterials migration in soil along with its components are indicators of danger of water migration. In general, effects of nanomaterials in soil matrix can be assessed by investigating bioaccumulation by roots and further translocation in hierarchy, bioavailability of nanomaterials for plants, soil microbe toxicity and migration depth along with other soil components. However, based on aggregative state of nanomaterial, bioavailability can be modulated. For instance, dissolved form of nanomaterials possesses higher bioavailability and consequently they are more toxic in soil structure. As aforementioned, nanomaterials toxicity assessment through investigating their bioavailability for plants is limited because the root exudates can influence nanomaterials distribution in rhizosphere. Therefore, we should consider soil structure and composition diversity, pH, alkaline status, cation exchange capacity, clay content, organic matter concentration and minerals concentration, during assessment.

With the allowance for complications in detecting nanomaterials in soil matrices, studies on soil organism have become an alternative strategy for investigating interaction. In general, soil contains natural colloids at low concentrations and due to variation in ionic strength nanomaterials tend to aggregate and sediment in soil. One of the problems in the study of NPs is related to the revelation of engineered NPs in the soil in the presence of natural colloids. Natural NPs in the soil (soil colloids) are difficult to separate and characterize.

The complexity of the soil organo-mineral composition and the unpredictable dynamics of soil properties in time and space may create problems in the structural and functional analysis of the biotic complex of soil under the impact of conventional pollutants, whose chemical transformations are well understood. It is known that behavior of NPs in natural media differs from that of coarser particles of the same material. Hence, nanomaterials can more easily enter into chemical reactions with other environmental components compared to coarser objects of the same composition; they are capable of forming complexes with previously unknown properties. An important factor for assessing nanomaterials impact on living organisms is the effect of the nanomaterial's interactions.

It was experimentally proven that soil can act as reliable filter for the migration of NPs if it contains high concentration of clay materials or has a high ionic strength. In addition, the filtration of NPs depends on the aqueous solution composition. It is interesting that the presence of humic acids or surfactants allows NPs to freely pass-through sand. Under hydroponic conditions, toxic effects of engineered nanomaterials on higher plants were frequently observed [28]. The leakage of nanomaterials into the environment, especially soil is one of the most serious threat to microbial communities in ecosystems.

Microorganisms in soil are essential to maintain soil function both in natural and agricultural soil due to their role in key processes, such as, decomposition of organic matters, structure formation, toxin removal, nutrient, such as, carbon, nitrogen, phosphorous, and sulfur recycle. In addition, they play key role in suppressing soilborne/plant diseases and promote plant growth. Changes in the composition and structure of soil microflora can be critical for the functional integrity of soil. Therefore, protection of environment and beneficial soil microorganisms from nanomaterials toxicity is essential, in spite of their beneficial role in plant development.

Nanomaterials comprise of key characteristics that are believed to exert essential controller on their environmental behavior, ecotoxicity, and fate. These characteristics are: physical properties, particle size, diameter, and shape and chemical properties (such as: charge around the particles, acid-base character of surface, surface charge, hydrophobicity, and hydrophilicity). These characteristics often define extent of transformation the NPs, which in turn modulates their fate, ecotoxicity, and behavior.

Nanomaterial transformation includes surface sorption, aggregation, sedimentation, and dissolution to the ionic metal. Soil represents a complex medium for investigating their physicochemical behavior of engineered nanomaterials. In comparison with the dissolved phase, in which behavior can be studied mostly in terms of particle stability against aggregation, soils are both solid matrix and contain appreciable number of colloidal particles with which nanomaterials may interact. In the background of ecotoxicity, a key problem is the understanding of how specific organism group is exposed to nanomaterials present in different phases, such as: soil or soil-water and their influence on microbial community.

In addition, nanomaterials behavior within soil matrix will be further complicated due to presence of soil components, such as, humic compounds or clay molecules as they possess charge around the particles and can influence aggregation of nanomaterials. Such soil components may also form colloids in aqueous phase and can interact with nanomaterials. For instance, titanium dioxide nanomaterial aggregation rate in soil suspension is negatively correlated with soil properties, such as: organic matter and clay concentration, whilst, positively correlated to ionic strength, pH, and zeta potential [42].

Pesticides are well-known chemical compounds in agriculture [26]. Nine out of 12 most harmful and unrelenting organic chemical compounds on earth have been acknowledged as pesticides and their derivatives. Because of pesticides strong recalcitrant property, conventional approaches lag behind to treat these hazardous chemicals. However, it is well professed that numerous factors, such as, light intensity, media, and humic content can govern the degradation of pesticides and their derivatives under ambient environment [25].

The exhaustive utilization of pesticides for controlling diseases, invasive plants and insects is necessary in agriculture practice to maximize the productivity and to balance the quality of final produce. The potential risk of pesticides to growing consumer health and concern regarding food safety and quality has evidenced the need to study the techniques capable of degrading these residuals in food. However, pesticides are intended to destroy and control the weeds and pests.

The intensive utilization of agrochemicals also demonstrated the adverse effect on environment, such as, threat to endangered birds and insect species, decreased population of insect pollination. In addition, on use of pesticide compounds, eye, skin, and nervous system related issues and cancer upon prolonged exposure have been detected [16]. Nano-technology-based approaches have engrossed worldwide consideration in recent times for the detection, degradation, and removal of hazardous pesticides. These approaches are known to be highly specific during detection and degradation of pesticides [31].

Intentional removal of pesticide residues necessitates processing techniques deliberately designed to remove pesticides and their degradation products. Such techniques would not normally be employed in food processing because they do not enhance the food value. Theoretically feasible methods are available for a number of situations, but, except for a few cases, little serious attention has been given to evaluating potentially effective processes.

Leaching and sediment transport by water redistribute soil pesticides and may, in a local sense, represent the removal. Vaporization, degradation by soil organisms or inorganic reactants, and uptake by plants are also important in soil pesticide dynamics. Recently instituted massive research and monitoring programs are now beginning to provide reliable data that define the magnitude of these various processes as they occur under diverse conditions of soil types and climates. This important information is providing few leads, however, that might be exploited for intentional removal of pesticides from soils [20].

Titanium dioxide has exceptional photo-catalytic activity at different pH and temperature, cost-effective, stability towards chemicals and non-toxicity. The band gap of titanium dioxide is 3.2 eV, which limits its photo catalytic activity only to the UV-region. However, dye sensitization and modification of surface of titanium dioxide NPs with different metal oxides, non-metals or carbon-based materials helps in extending the photo-catalytic activity of NPs to the visible region. These TiO_2 NPs have been used to extract and degrade pesticides from various samples [39].

Water could conceivably be cleaned up by procedures for removal of the sediments carrying adsorbed pesticides and by use of activated media to absorb the free compounds. Sedimentation processes already employed in purification of municipal waters undoubtedly serve this function to some degree [2]. Activated carbon will remove organochlorine pesticides from water, but large quantities of carbon are required to achieve maximum removals.

The oxidants, ozone, and potassium permanganate ($KMnO_4$) have not produced promising results relative to removal of organochlorine residues. Biological trapping of residues has also been suggested for water purification. This probably occurs in sewage lagoons through the growth of algae, but the algae would have to be harvested and destroyed elsewhere to achieve actual removal. Ion exchange resins could be used to reduce the content of ionic pesticides, while solvent extraction techniques could remove non-ionic compounds. All of these processes require extensive engineering and chemical evaluation for proper assessment [6, 20].

3.4 SUMMARY

The potential application of nanotechnology in agriculture includes development of nano-fertilizers for enhanced productivity, nano-based pesticide formulations for plant protection and disease management, nano-enabled sensors for plant and soil health monitoring. Presence of nanomaterials in any environment or matrix highly influence their physical and chemical characteristics and understanding their biological fate and translocation in soil matrices is highly challenging and possess complexity in soil-plant interaction platforms. The broad range of nanomaterials have been utilized as micro- and macronutrients to enhance soil fertility, plant productivity and nutrients uptake rate. To enhance micronutrients bioavailability for plants, nano-encapsulated micronutrients formulation or nano-minerals supplementation offers a better strategical platform.

KEYWORDS

- nano-fertilizers
- nano-nutrition
- nano-pesticides
- nanoscience
- pesticide removal
- plant development
- titanium dioxide

REFERENCES

1. Achari, G. A., & Kowshik, M., (2018). Recent developments on nanotechnology in agriculture: Plant mineral nutrition, health, and interactions with soil microflora. *Journal of Agricultural and Food Chemistry, 66*(33), 8647–8661.

2. Bajwa, U., & Sandhu, K. S., (2014). effect of handling and processing on pesticide residues in food: A review. *Journal of Food Science and Technology, 51*(2), 201–220.

3. Bin-Hussein, M. Z., Zainal, Z., Yahaya, A. H., & Foo, D. W. V., (2002). controlled release of a plant growth regulator, A-naphthaleneacetate from the lamella of Zn-Al-layered double hydroxide nanocomposite. *Journal of Controlled Release, 82*(2), 417–427.

4. Birwal, P., Rangi, P., & Ravindra, M. R., (2019). Nanotechnology applications in packaging of dairy and meat products. Chapter 1. In: Lohith, K. D. H., Goyal, M. R., & Suleria, H. A. R., (eds.), *Nanotechnology Applications in Dairy Science: Packaging, Processing, Preservation* (pp. 3–26). Oakville - ON: Apple Academic Press.

5. Biswas, S., Pandey, N., & Rajik, M., (2012). Inductions of defense response in tomato against fusarium wilt through inorganic chemicals as inducers. *Plant Pathology and Microbiology, 3*(4), 3–4.

6. Chen, C., Qian, Y., Chen, Q., Tao, C., Li, C., & Li, Y., (2011). Evaluation of pesticide residues in fruits and vegetables from Xiamen, China. *Food Control, 22*(7), 1114–1120.

7. Chhipa, H., (2016). Nanofertilizers and nanopesticides for agriculture. *Environmental Chemistry Letters, 15*(1), 1–8.

8. Corradini, E., De Moura, M., & Mattoso, L., (2010). Preliminary study on the incorporation of NPK fertilizer into chitosan nanoparticles. *Express Polymer Letters, 4*(8), 509–515.

9. Correia, A. A. S., & Rasteiro, M. G., (2016). Nanotechnology applied to chemical soil stabilization. *Procedia Engineering, 143*(1), 1252–1259.

10. Dasarahalli-Huligowda, L. K., Goyal, M. R., & Suleria, H. A. R., (2019). *Nanotechnology Applications in Dairy Science: Packaging, Processing, and Preservation* (1st edn., p. 275). Oakville - ON: Apple Academic Press.

11. Duhan, J. S., Kumar, R., Kumar, N., Kaur, P., Nehra, K., & Duhan, S., (2017). Nanotechnology: The new perspective in precision agriculture. *Biotechnology Reports, 15*(1), 11–23.

12. El-Kereti, M., El-Feky, S., Khater, M., Osman, Y., & El-sherbini, E., (2013). ZnO nanofertilizer and He Ne laser irradiation for promoting growth and yield of sweet basil. *Recent Pat. Food Nutrit., 5*(3), 169–181.

13. Hernandez-Viezcas, J. A., Castillo-Michel, H., Andrews, J. C., Cotte, M., Rico, C., Peralta-Videa, J. R., Ge, Y., et al., (2013). *In Situ* synchrotron X-ray fluorescence mapping and speciation of CeO_2 and ZnO nanoparticles in soil cultivated soybean (*Glycine max*). *ACS Nano, 7*(2), 1415–1423.

14. Hossain, K. Z., Monreal, C. M., & Sayari, A., (2008). Adsorption of urease on PE-MCM-41 and its catalytic effect on hydrolysis of urea. *Colloids and Surfaces B: Biointerfaces, 62*(1), 42–50.

15. Ibrahim, R. K., Hayyan, M., Al-Saadi, M. A., Hayyan, A., & Ibrahim, S., (2016). Environmental application of nanotechnology: Air, soil, and water. *Environmental Science Pollution Research, 23*(14), 13754–13788.

16. Ikeura, H., Kobayashi, F., & Tamaki, M., (2011). Removal of residual pesticides in vegetables using ozone microbubbles. *Journal of Hazardous Materials, 186*(1), 956–959.

17. Janmohammadi, M., Amanzadeh, T., Sabaghnia, N., & Dashti, S., (2016). Impact of foliar application of nano micronutrient fertilizers and titanium dioxide nanoparticles on the growth and yield components of barley under supplemental irrigation. *Acta Agriculturae Slovenica, 107*(2), 265–276.

18. Kabiri, S., Degryse, F., Tran, D. N. H., Da Silva, R. C., McLaughlin, M. J., & Losic, D., (2017). Graphene oxide: A new carrier for slow release of plant micronutrients. *ACS Applied Materials & Interfaces, 9*(49), 43325–43335.

19. Kang, H., Hwang, Y. G., Lee, T. G., Jin, C. R., Cho, C. H., Jeong, H. Y., & Kim, D. O., (2016). Use of gold nanoparticle fertilizer enhances ginsenoside contents and anti-inflammatory effects of red ginseng. *Journal of Microbiology and Biotechnology, 26*(10), 1668–1674.

20. Keikotlhaile, B. M., Spanoghe, P., & Steurbaut, W., (2010). Effects of food processing on pesticide residues in fruits and vegetables: A meta-analysis approach. *Food and Chemical Toxicology, 48*(1), 1–6.

21. Kumar, D. L., Mitra, J., & Roopa, S., (2020). Nanoencapsulation of food carotenoids. Chapter 7. In: Dasgupta, N., Ranjan, S. and Lichtfouse, E. (eds.), *Environmental Nanotechnology* (Vol. 3, pp. 203–242). London: Springer.

22. Kumar, D. L., & Sarkar, P., (2017). Nanoemulsions for nutrient delivery in food. Chapter 4. In: Ranjan, S., Dasgupta, N., & Lichtfouse, E., (eds.), *Nanoscience in Food and Agriculture* (pp. 81–121). Cham: Springer.

23. Lin, D., & Xing, B., (2007). Phytotoxicity of nanoparticles: Inhibition of seed germination and root growth. *Environ. Pollut., 150*(2), 243–250.

24. Lohith-Kumar, D. H., & Sarkar, P., (2018). Encapsulation of bioactive compounds using nanoemulsions. *Environmental Chemistry Letters, 16*(1), 59–70.

25. Lv, T., Zhang, Y., Zhang, L., Carvalho, P. N., Arias, C. A., & Brix, H., (2016). Removal of the pesticides imazalil and tebuconazole in saturated constructed wetland mesocosms. *Water Research, 91*, 126–136.

26. Lyu, T., Zhang, L., Xu, X., Arias, C. A., Brix, H., & Carvalho, P. N., (2018). Removal of the pesticide tebuconazole in constructed wetlands: Design comparison, influencing factors and modeling. *Environmental Pollution, 233*, 71–80.

27. Mani, P. K., & Mondal, S., (2016). Agri-nanotechniques for plant availability of nutrients. Chapter 11. In: Khodakovskaya, D. S., Chittaranjan K., & Mariya, V. K., (eds.), *Plant Nanotechnology* (pp. 263–303). Cham: Springer.

28. Meyer, J. N., Lord, C. A., Yang, X. Y., Turner, E. A., Badireddy, A. R., Marinakos, S. M., Chilkoti, A., et al., (2010). Intracellular uptake and associated toxicity of silver nanoparticles in *Caenorhabditis elegans*. *Aquat. Toxicol., 100*(2), 140–150.

29. Pandey, A. C., Sanjay, S. S., & Yadav, R. S., (2010). Application of ZnO nanoparticles in influencing the growth rate of *Cicer arietinum*. *Journal of Experimental Nanoscience, 5*(6), 488–497.

30. Prasad, T., Sudhakar, P., Sreenivasulu, Y., Latha, P., Munaswamy, V., Reddy, K. R., Sreeprasad, T., et al., (2012). Effect of nanoscale zinc oxide particles on the germination, growth and yield of peanut. *Journal of Plant Nutrition, 35*(6), 905–927.

31. Rai, M., & Ingle, A., (2012). Role of nanotechnology in agriculture with special reference to management of insect pests. *Applied Microbiology and Biotechnology, 94*(2), 287–293.
32. Ramkumar, C., Vishwanatha, A., & Saini, R., (2019). Regulatory aspects of nanotechnology for food industry. Chapter 7. In: Lohith, K. D. H., Goyal, M. R., & Suleria, H. A. R., (eds.), *Nanotechnology Applications in Dairy Science: Packaging, Processing, Preservation* (pp. 169–184). Oakville - ON: Apple Academic Press.
33. Rengel, Z., (2015). Availability of Mn, Zn and Fe in the rhizosphere. *Journal of Soil Science and Plant Nutrition, 15*(2), 397–409.
34. Sanzari, I., Leone, A., & Ambrosone, A., (2019). Nanotechnology in plant science: To make a long story short. *Frontiers in Bioengineering, 7*(1), 1–12.
35. Shang, Y., Hasan, M., Ahammed, G. J., Li, M., Yin, H., & Zhou, J., (2019). Applications of nanotechnology in plant growth and crop protection: A review. *Molecules, 24*(14), 2558–2563.
36. Sharma, P., Bhatt, D., Zaidi, M. G. H., Saradhi, P. P., Khanna, P. K., & Arora, S., (2012). Silver nanoparticle-mediated enhancement in growth and antioxidant status of *Brassica juncea. Applied Biochemistry and Biotechnology, 167*(8), 2225–2233.
37. Shivaram, S. H., & Saini, R., (2019). Spray drying-assisted fabrication of passive nanostructures: From milk protein. Chapter 3. In: Lohith, K. D. H., Goyal, M. R., & Suleria, H. A. R., (eds.), *Nanotechnology Applications in Dairy Science: Packaging, Processing* (pp. 45–68). Oakville - ON: Apple Academic Press.
38. Singh, N., Amist, N., Yadav, K., Singh, D., Pandey, J., & Singh, S., (2013). Zinc oxide nanoparticles as fertilizer for the germination, growth and metabolism of vegetable crops. *Journal of Nanoengineering and Nanomanufacturing, 3*(4), 353–364.
39. Sun, H., Zhang, X., Niu, Q., Chen, Y., & Crittenden, J. C., (2007). Enhanced accumulation of arsenatein carp in the presence of titanium dioxide nanoparticles. *Water, Air, and Soil Pollution, 178*(4), 245–254.
40. Thul, S., Sarangi, B., & Pandey, R., (2013). Nanotechnology in agroecosystem: Implications on plant productivity and its soil environment. *Expert Opinion on Environmental Biology, 2*(1), 1–7.
41. Thul, S. T., & Sarangi, B. K., (2015). Implications of nanotechnology on plant productivity and its rhizospheric environment. Chapter 3. In: Siddiqui, M H., Al-Whaibi, M. H., Mohammad, F. (eds.), *Nanotechnology and Plant Sciences* (pp. 37–53). Cham: Springer.
42. Tourinho, P. S., Van, G. C. A., Lofts, S., Svendsen, C., Soares, A. M., & Loureiro, S., (2012). Metal-based nanoparticles in soil: Fate, behavior, and effects on soil invertebrates. *Environ. Toxicol. Chem., 31*(8), 1679–1692.
43. Wang, P., Lombi, E., Zhao, F. J., & Kopittke, P. M., (2016). Nanotechnology: A new opportunity in plant sciences. *Trends in Plant Science, 21*(8), 699–712.
44. Xiong, T., Dumat, C., Dappe, V., Vezin, H., Schreck, E., Shahid, M., Pierart, A., & Sobanska, S., (2017). Copper oxide nanoparticle foliar uptake, phytotoxicity, and consequences for sustainable urban agriculture. *Environmental Science & Technology, 51*(9), 5242–5251.
45. Yang, F., Liu, C., Gao, F., Su, M., Wu, X., Zheng, L., Hong, F., & Yang, P., (2007). The Improvement of spinach growth by nano-anatase TiO_2 treatment is related to nitrogen photoreduction. *Biological Trace Element Research, 119*(1), 77–88.

46. Yuvaraj, M., & Subramanian, K., (2015). Controlled-release fertilizer of zinc encapsulated by a manganese hollow core shell. *Soil Science & Plant Nutrition, 61*(2), 319–326.

47. Zhao, L., Sun, Y., Hernandez-Viezcas, J. A., Servin, A. D., Hong, J., Niu, G., Peralta-Videa, J. R., et al., (2013). Influence of CeO_2 and ZnO nanoparticles on cucumber physiological markers and bioaccumulation of Ce and Zn: A life cycle study. *Journal of Agricultural and Food Chemistry, 61*(49), 11945–11951.

48. Zheng, L., Zhou, M., Chai, Q., Parrish, J., Xue, D., Patrick, S. M., Turchi, J. J., et al., (2005). Novel function of the flap endonuclease 1 complex in processing stalled DNA replication forks. *EMBO Reports, 6*(1), 83–89.

PART II

Role of Nanotechnology in Food Products, Nutraceuticals, and Therapeutics

CHAPTER 4

NANOENCAPSULATION METHODS IN IRON FORTIFICATION OF DAIRY FOOD MATRICES

AMRITA POONIA

ABSTRACT

Dairy products are good medium for fortifying with iron and other essential minerals and nutrients due to their high bioavailability and stability during digestion. The stability of nutrients depends on temperature, oxygen concentration, pH, light, enzymes, and different agents, which cause oxidizing effect. However, direct addition of iron in dairy products decreases the bioavailability due to reaction of milk fat and protein with iron causing development of unpleasant color and flavor. To solve these issues, nanoencapsulation of essential minerals is a successful approach to increase the shelf-life and sensory properties.

4.1   INTRODUCTION

Milk is considered as a complete food and milk-derived products are major sources of essential nutrients required for our body. Milk contains approximately 5% lactose, 3% protein, 4% lipid and 0.7% mineral salts [33]. Dairy products are poor sources of iron, copper, and zinc, and milk contributes little to the total iron and zinc intake. Many factors affect the bioavailability of these nutrients and peoples do not receive their daily dietary needs of such nutrients. Processing of dairy products during heating, pasteurization, homogenization, ultra-heat treatment (UHT) and spray drying causes loss of certain nutrients to some extent [7, 56]. Hence, to recover the lost nutrients and adding certain health beneficial bioactive components, fortification is an effective way devoid of causing any change in the nutritional patterns [44].

The concept of fortification involves the addition of micronutrients to balance the destruction caused through processing methodologies. Dairy products are consumed by people of all ages for multiple purposes. Children consume milk during their developmental stage; while adults consume yogurt, cheese, and ice cream for nutrition, pleasure, and/or taste. In women, dairy products help in prevention of osteoporosis and other bone- and joint-related diseases. Thus, dairy products are effective vehicles for fortification, as they are consumed by a broad spectrum of the global population [21]. This ensures the nutrients uniformly to reach a wide group of population.

Minerals are major part of micro- and macro nutrients associated with biological functions and cellular metabolism. Though, animals, and plants are unable to synthesize the minerals that are absorbed by plants absorb *via* plant roots. Hence, major dietary sources of minerals are fruits and vegetables. The high risk of minerals prevails during different stages of human growth, such as: adolescence, pregnancy; while, dietary restrictions (such as: low-calorie diets, vegetarians, and geriatric patients) explore the importance of minerals in the balanced diet. Recently nanotechnology has been introduced to the food industry at various processing operations. Nano-sensors, tracking devices, targeted nutrient delivery systems, food packaging, food product development and accuracy in processing are such examples [22].

Milk proteins, carbohydrates, and fat molecules have been used after modifications in the food packaging. These help to improve the sensory properties including texture modification. Use of nanoparticles (NPs) in designing delivery vehicles has been studied for selected bioactive compounds and their absorption. Nutraceuticals as compared to other ingredients become more stable in food structures. In addition, they also help in generating new ideas of product development and the extension of their stability. NPs are also integral part of sensors used in tracing foods, spoilage with antimicrobial agents to protect the food from food borne pathogens and concentration of liquid food products [29–31, 45].

The consistency of nanomaterials used in the food sector depends on the storage environment conditions, such as: temperature, oxygen availability and pH. For instance, nano-capsules are greasy or hydrophilic hollow space bounded by thin wall substance, and their stability varies with pH, temperature, and ionic strength. Different types of wall materials (such as: lipids, surfactants, biopolymers, and chemical polymers) may be used during the manufacturing of nanocapsules [13, 14, 29, 49].

Nanoemulsions, NPs, nanosuspensions, nanostructured lipid carriers, nanoclusters, polysaccharides, nano-clays, nanosized liposomes, proteins micelles, biopolymer NPs, complexes/conjugates of polysaccharides are nanomaterials that has been utilized in dairy sector for multiple purposes. Ingredients used for preparing nanoemulsions are lipophilic phase, hydrophilic phase, and amphiphilic stabilizers.

The most common considered delivery systems are colloidal form of nanoliposomes (NLCs). Phospholipids, cholesterol, phenolic compounds, and solvents are the ingredients used for the preparation of NLCs. Ingredients used for the preparation of solid liquid nanoparticles (SLN) and NLCs are liquid oil, solid lipid, and stabilizers. The high-pressure homogenization, ultrasonication, electro-spinning, and co-acervation methods are used for the preparation of NPs.

This chapter emphasize the nano-fortification of minerals in dairy food products to lessen the organoleptic troubles and to enhance the bioavailability. Various problems that occur during the minerals fortification and different encapsulation strategies are also discussed in this chapter. Safety and toxicity issues of NPs, various organizations working in this sector have also been included.

4.2 PREREQUISITES FOR FORTIFICATION FOR DAIRY FOODS

Milk as carrier matrix is one of nutrient rich foods required for human body. Milk and milk products are consumed by the people of all age groups around the world. The trends in food fortification fluctuate depending on the nutrient requirements of the people. The prerequisites of food fortification are: constant demand for food and unaffected by fortification, less adverse-effect on taste, texture, appearance, and odor of food matrices, higher bioavailability of fortified nutrient in gastro-intestinal tract and should possess a provable enhanced health effect [11].

Through fortification of minerals and nutrient, the nutritive value of milk and milk products can be enhanced. Processing parameters and the environmental conditions affect the bioactive compounds. The stability under gastric conditions are major concerns encountered during design of encapsulation matrix for food applications. In addition, the stability of nutrients depends on temperature, oxygen concentration, pH, light, enzymes, and different components present in food matrix during storage [32].

Direct mixing of iron or other minerals into dairy based products has many constraints, such as: low bioavailability due to untargeted absorption in digestive tract, development of off-flavor due to oxidation and textural deformation due to reaction of iron with milk proteins [15]. Therefore, to enhance the organoleptic properties of food matrix along with bioavailability, nanoencapsulation strategies could be an alternative approach [26].

4.3   DAIRY PRODUCTS FORTIFIED WITH IRON

Human milk contains iron 6.8 mmol L^{-1} and 9.3 mmol L^{-1} in milk and colostrum, respectively in a complex form with lactoferrin [50]. Lactoferrin is a glycoprotein that plays an important role in the human innate immune system. In human milk, lactoferrin is an iron-free apoprotein that can bind two ferric ions. Human milk contains citrate, which facilitates lactoferrin binding to iron. It is the most dominant whey protein in human milk. Lactoferrin is present in high quantities in colostrum and declines over time. Lactoferrin is resistant to proteolysis, especially in its iron-saturated form. Lactoferrin has high affinity towards iron and increases in the weak acidic medium.

The benefit with iron fortification in milk formulas has not been proven clinically. There is controversy with efficiency of fortification of iron in infant foods. Hence, lactoferrin is purified and used with iron. The fortification of iron with lactoferrin is more costly as compared to fortification with salts of iron. However, lactoferrin in the isolated form is frequently used as supplementation in infant foods [9].

Various iron salts used in enrichment of food products are listed in Table 4.1. Iron salts used should have high solubility, zero precipitation and no modification in pH. Citrates and phosphates are some examples with good buffering capacity. Ferrous chloride is rapidly oxidized immediately after solubilization. Various minerals used for mineral fortification in dairy-based products are given in Table 4.2. Use of various technologies has been reported for preparation of encapsulated minerals for fortification of dairy products. Different encapsulation technologies for fortification of nanoencapsulated minerals in dairy based products are tabulated in Table 4.3.

TABLE 4.1 Different Forms of Iron Salts with Their Solubility and Bioavailability

Different Forms of Fe Salt	Water Solubility	Bioavailability	Sensitivity to Oxidation
Ferrous chloride	NA	50	NA
Ferrous citrate	NA	74	NA
Ferrous fumarate	Low	100	High
Ferrous lactate	NA	89–106	NA
Ferrous phosphate Insoluble	Insoluble	27	Comparatively low
Ferrous pyrophosphate	NA	30	NA
Ferrous succinate	NA	92	NA
Ferrous sulfate	Soluble	100	Very high
Ferrous tartrate	NA	62	NA

TABLE 4.2 Mineral Compounds used for Fortification of Dairy Products

Mineral Compounds	State of Mineral Used	Solubility in Water	References
Mineral protein complexes	Whey proteins, casein, casein phospho-peptide (CPP) and CPPs-Zn complexes	Fortification of cheese, high moisture dairy foods	[15]
Mineral salts (soluble)	Fe^{2+}, Fe^{3+}, Zn^{2+}, Zn^{3+}	Soluble	[15]
Minerals (element)	Reduction of salts with H_2 or CO, electrolysis, and the Mond process	Insoluble/partially soluble	[38]

4.4 ISSUES RELATED TO IRON FORTIFICATION IN DAIRY PRODUCTS

Fortification of iron in dairy products has multiple issues especially in the dairy desserts as compared to fermented dairy products due to the oxidized off-flavor, changes in color and metal, e.g., flavor probably due to lipid peroxidation of milk fat. In photooxidation coloring agents induce

TABLE 4.3 Fortification of Nanoencapsulated Minerals in Dairy-based Products Using Different Encapsulation Technologies

Product	Minerals	Technology/Salt	Size (μ)/ Mineral Content	EE	References
Cheese	Fe	FAE	2–5	72	[33]
Cheese	Calcium	Calcium chloride	0.02–0.03 (0.05%)	–	[54]
Dairy products	Mg	Emulsification	9–10	>99	[8]
Beverage yogurt	Fe	FAE	2–5	75.0	[27]
Edam cheese	Zinc	$ZnSO_4$, ZnO, zinc acetate, $ZnCl_2$	NA	NA	[2]
Milk	Fe	Liposome	NA	81.3	[1]
Milk	Fe	Liposome	NA	62.9–74.8	[18]
Milk	Fe	Fatty acid esters (FAE)	NA	85.0	[1]
Milk	Fe	FAE	NA	62.9–74.8	[18]
Milk	Fe	FAE	2–5	75.0	[34]
Milk	Fe	Modified solvent evaporation	15.54	91.58	[19]
Milk	Fe	Emulsification	NA	62.9–74.8	[18]
Milk	Fe	Emulsification	2–12	93.63	[10]
Milk	Fe	Freeze-drying	NA	62.9–74.8	[18]
Milk	Calcium	Ca chloride dihydrate, Calcium L-lactate tetrahydrate, calcium gluconate monohydrate, $CaCO_3$	(50–75–100 mg/100 mL)	NA	[51]
Milk	Selenium	NA	NA	NA	[6]
Soy milk	Ca	Emulsification	NA	91.18	[43]

TABLE 4.3 *(Continued)*

Product	Minerals	Technology/Salt	Size (μ)/ Mineral Content	EE	References
Turkish white cheese	Se and Zn	NA	NA	NA	[17]
Yoghurt	Fe	Niosome	1.44, 7.21	72–84	[20]
Yoghurt	Fe	Niosome (supercritical CO^2)	NA	25.1	[55]
Yoghurt	Fe	Emulsification	NA	NA	[53]
Yoghurt	Fe	Salt induced cold gelation	1	82.0	[37]
Yoghurt	Calcium	Calcium lactate, calcium gluconate	400–600–800 mg/100 g of calcium lactate 600–800–1,000 mg of calcium gluconate	NA	[39]

Abbreviations: AG: Arabic gum; HPMC: hydroxypropyl methylcellulose; LMP: low methoxyl pectin; MD: maltodextrin; MG: mesquite gum; MS: modified starch; PGMS: polyglycerol monostearate; WPC: whey protein concentrate; and WPI: whey protein isolate.

Note: EE is based on mg/kg.

peroxidation and iron produce black color due to change in valence. Use of ferrous salts results in bitter taste and metallic flavor.

Stability of mineral fortification depends on their stability in food and possible interactions with milk proteins. Minerals are less susceptible as compared to vitamins towards the physical and chemical factors. In addition, minerals are less stable when exposed to heat, air, or light. Minerals fortification does not influence the sensory and physicochemical properties of milk and milk products.

4.5 NANOENCAPSULATION TECHNIQUES

On the basis of mechanism and basics ingredients used nanoencapsulation technologies are divided into five major groups as shown in Figure 4.1.

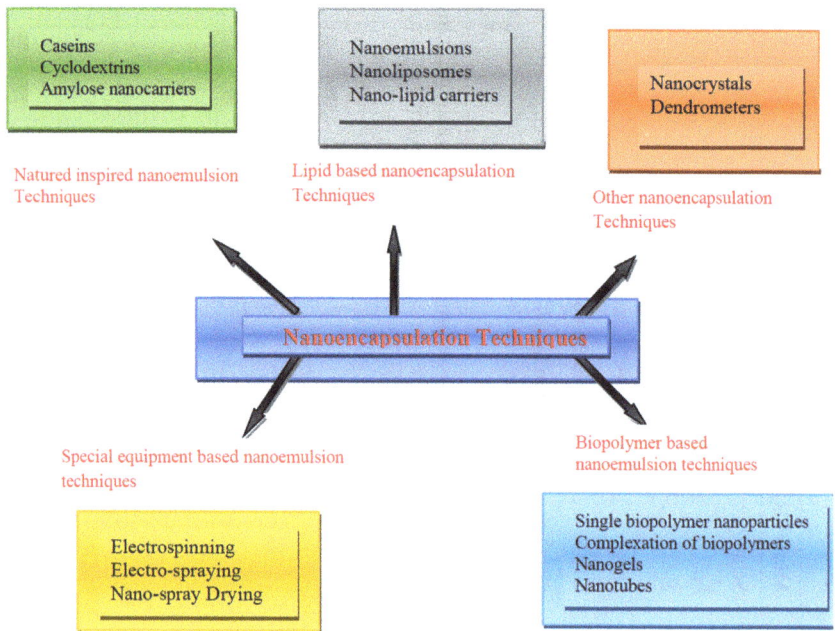

FIGURE 4.1   Nano encapsulation technologies for bioactive ingredients.

Various encapsulation techniques are used to entrap the bioactive compounds that are used for the dairy based products. The applicability of any procedure depends on the ingredient need to be encapsulated and the kind of the product carrier. Nano-functional ingredients in form of the nanoemulsion and nano-powder has been used in dairy products. In recent research studies, the addition of powdered form of functional ingredients has been reported in dairy products.

4.5.1   METHODS FOR NANOENCAPSULATION

4.5.5.1   ELECTRO-SPRAYING

Electro-spraying is a simple and effective method of nano encapsulation, where yield of nano-capsules depends on the environmental factors, such as: pressure and temperature, pH. Using food grade polymeric particles, electro-spraying can be used to produce micro to nano scale particles. Basic modes of electro-spraying of viscous polymer solutions are: dripping based mode, an alternating ejection mode, and single cone stream that has stable Taylor shape cone [24].

4.5.5.2   ELECTROSPINNING

There are three basic components of electrospinning, such as: (1) high voltage-based supply; (2) spinneret that is made of a blunt type needle syringe; and (3) a metal-based screen for collection purpose. The Kv of direct current is supplied to induce an additional electric charge in liquid form of the polymer. Ghorani and Tucker [16] studied mechanism of electrospinning and electrospun nanofibers to carry bioactive components. Bioactive components of these types can considerably control the mechanical properties and functionality of fiber-based products.

Spasova et al. [52] used different types of polymers and solvents to produce diverse forms of electrospun fiber cross sections. Temperature and humidity can affect the structure of fiber and the output of electrospinning process.

4.5.5.3   NANO-SPRAY DRYING

Spraying consists of generation of particulate dried powders from the fluid in a hot drying medium. There are three basic steps in spray drying:

droplets are generated; drying of the heating gas, drying of solution droplets; and the collection of the particles. The single step process is used for dry powder formation of aqueous, emulsions, organic solutions, and suspensions. Nano spray drying is mainly used in pharmaceutical, chemical, and food industry. This technique is simple and fast as compared to other techniques. The cost of spray dried NPs is less as compared to freeze drying. The main drawback of spray dryers for producing NPs is the partial separation efficiency and collection of the particles in submicron size using cyclone type separator [46].

The cyclones with typical types are not able to collect the particle size < 2 μm. Another drawback is turbulent type flow of gas inside chamber used for drying, which causes the deposition of the particles on the wall of the chamber. In addition, traditional atomizers do not permit the fine particles being produced to attain the submicron size of the particle. Recently some progress has been reported for spray dryers, which can produce particle size of < 1 μm.

In 2009, Nano type spray dryer model B-90 was developed and used for the production of submicron scale particles using spray drying. This model of spray dryer was basically for the laboratory purpose. This spray dryer could produce the particles of submicron size using nanoemulsions and nanosuspension. The unique features of this technique are its extremely well-organized electrostatic type precipitator for collection of NPs, formation of fine droplets and the mild stream of the laminar gas used for drying.

4.6   APPLICATIONS OF NANO-MATERIALS IN DAIRY PRODUCTS

Isoflavone, conjugated linoleic acid (CLA), chitosan, mistletoe, *Inonotus obliquus* and the extract from the sprouts of peanuts are the main nutraceutical compounds that are used for fortification of dairy products. The utilization of encapsulated nutrients for fortifying dairy-based products is a beneficial method of attaining more nutrient intake. Fortification of dairy based products with encapsulated form of nutrients is shown in Figure 4.2.

4.6.1   FUNCTIONAL MILK

Functional milk and milk products are well-known for their health benefits to us. Use of various food raw materials has been considered to prepare the functional milk. Various studies have been reported for production of functional milk by use of some new nano-powdered bioactive type compounds. Utilization of nano functional ingredients in milk products is shown in Table 4.4.

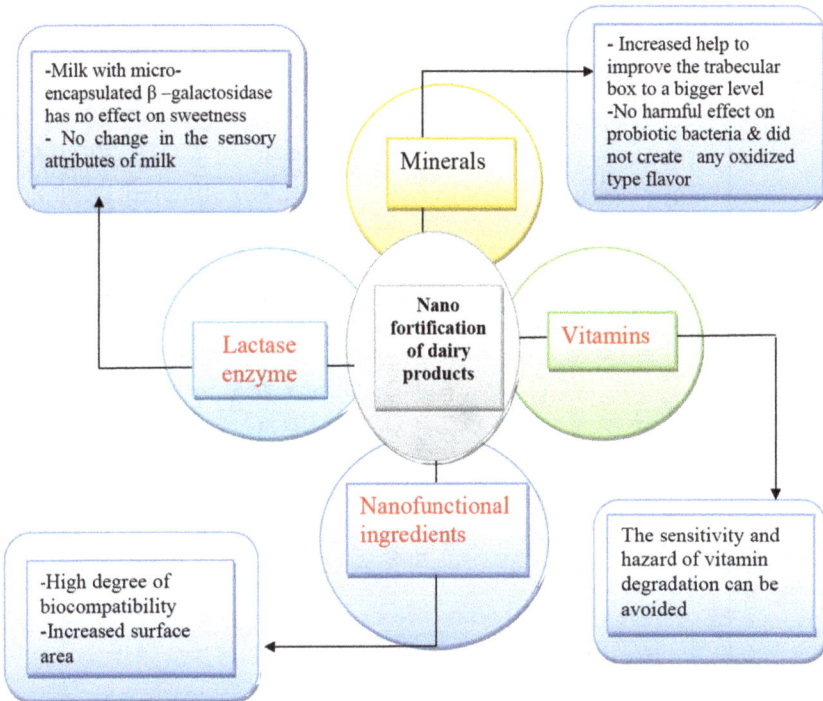

FIGURE 4.2   Fortification of dairy products with encapsulated nutrients.

Resinack and Muller [41] compared nano-powdered chitosan (NPC) and powdered chitosan for their biological activity and they reported that NPC shows elevated level of biological activities than the powdered form. No change was reported in the pH and titratable acidity after addition of NPC. Use of NPC in concentrations @ 1–3% v/v in milk resulted in a better product, but elevated levels of NPC provided more astringency, which adversely affects the milk quality. Seo et al. [47, 48] studied addition of soluble type ascorbic acid NPs for enhancing the functionality of milk. They reported the enhancement of the levels of ascorbic acid in milk, which resulted in several benefits to human health.

TABLE 4.4 List of Dairy Products Using Nano-Functional Ingredients

Dairy Products	Nano-Functional Ingredients	Remarks	References
Asiago cheeses	Nano-powdered red ginseng	Asiago cheeses are acceptable, when prepared by addition of nano-powdered red ginseng	[12]
Caciocavallo cheese	Nano-peanut sprout	Addition of peanut sprouts in nano-powder form (NPS) in cheese improved the resveratrol through ripening	[28]
Cheese	Liposome-encapsulated enzyme including flavor zyme acid fungal protease, neutral bacterial protease, and lipase	This cocktail helps in balance ripening	[25]
Yoghurt	Nano peanut sprouts	Comparatively nano-powdered peanut sprout (NPS) added yogurt results in decreased pH as compares to the powdered form of peanut sprout	[3]
Yoghurt	Nano peanut sprouts	Use of NPPS (nano-powdered pea nut sprout) equal to 0.2% in yogurt results in increased antioxidant activity	[4]
Yoghurt	Nano-chitosan	Use of nano-powdered chitosan (NPC) @ 0.3–0.5% (v/v) could be incorporated in yogurt without any adverse effect on the pH, LAB, viscosity, and sensory attributes	[47]
Yoghurt	Nano-ginseng	Addition of nano-ginseng equal to 0.3% (w/v) added to functional yogurt is acceptable	[35]
Yoghurt	Nano powdered eggshell	Nano-powdered eggshell supplemented (NPES) yogurt is more stable as compared to powdered eggshell-supplemented yogurts during storage studies	[5]

4.7 SAFETY AND TOXICITY OF NANOPARTICLES (NPS) USED IN DAIRY INDUSTRY

Before the production processes of the nano materials, the possible toxicity and health risks should be characterized. NPs enter inside the body through skin penetration, breathing, and through mouth. As the NPs enters in the GIT, they stay inaccessible from blood by the epithelial cells. The absorption of NPs is higher in small intestine as compared to buccal cavity. Due to the accumulation of NPs inside cells of intestine, inflammatory bowel syndrome has been reported in rats [42].

The physicochemical alterations caused due to pH, ionic power, carbohydrates, lipids, and the amount of protein present inside body fluids are possible risk of toxicity of the digested NPs. After entering in the blood flow, NPs act in response with blood cells and proteins present in plasma and then enters the cells and modify the DNA present in cells and thus change their performance. Due to slow absorption of NPs inside body and bioaccumulation, it can result in toxicity. There is a need to understand the interaction between NPs and biological system and toxicological effects. There is also need to manufacture equipment and procedures for the evaluation of contact period rate of nanomaterials in the diverse media [40]. Promotion of new methods for testing toxicity of the nanomaterials and to set up analytical methods using computer software and the programming to decide about environmental and public health is need of the hour [36].

4.8 RESEARCH SECTORS FOR NANO-FOODS

Fortification of food is one of the well-known strategies to get better health and nutrition rank. There is a need to determine that whether the food product is safe with the existing nanomaterial used or it need some changes in order to consider the safety of functionality of food products. The literature review revealed safe applications of nano materials and their long-standing influence on the body. The ZnO and Ag as nanomaterials are usually considered for consumption of human beings with low/no toxicity.

Encapsulated nanomaterials (such as: asbestos) are unlike from non-toxic materials in the form and ability of fibers to stay in the biological environment. In totality, nano food products should include a process for safety assessment, in which phase studies concerning the contact, toxicity have to be conducted to obtain consistent information on the safety of the consumers. *In-vitro* and *in-vivo* tests are of immense significance to know any kind of toxicity due to consumption of nanomaterials in the cells and the living beings.

US Food and Drug Administration (FDA) recently performed some *in-vivo* studies to find out the impact of nanomaterials on living systems. FDA is also organizing *in-vitro* tests to assess the genotoxicity of these minute materials and to identify their probability of mutagenicity or carcinogenicity relating titanium dioxide and silver. Number of government agencies and private organizations (Table 4.5) are studying the safety aspects of nano food materials. At last, the regulatory points legislated for nano-based food products are still in initial stage and have to be finished by the collaboration of all the food regulatory bodies around the world [23].

TABLE 4.5 Governmental and Private Sectors Researching Various Aspects of Food Nanotechnology in India and Other Countries

Name of Institutions and Research Centers
Indian Institutes
Center for Nanotechnology and Advanced Biomaterials (CeNTAB), Thanjavur
Center of Nanotechnology, Indian Institute of Technology, Roorkee
Institute of Nanoscience and Technology, Mohali, Punjab
School of Nanoscience and Technology (NITC), Calicut
Other Countries
Center for Nanotechnology
Center for Nanotechnology, NASA
Center for Science Communication, University of Chester
European Nanotechnology Gateway
Food and Agriculture Organization of the United Nations
Food Nanotechnology Research Group, University of Chester
French Research Network in Micro and Nano Technologies, France
Functional Foods Research Center, University of Chester
German Federal Institute for Risk Assessment
Hydrocolloids Research Center, University of Chester
Institute of Food Research, UK
Iran Nanotechnology Initiative Council
Korean Food and Drug Administration
National Food Research Institute, NARO
National Institute for the Physics of Matter, Italy
National Nutrition and Food Technology Research Institute, Iran
National Research Council, Canada
National Science Foundation, United States

TABLE 4.5 *(Continued)*

Restricted Diets Research Center, University of Chester
United States Department of Agriculture National Institute of Food and Agriculture
US Food and Drug Administration
Public Sectors
Nanostructural Analysis Network Organization, Major National
Nanowork
NoWFOOD (Small and medium size food business)
Project on Emerging Nanotechnologies

4.9 SUMMARY

Fortification of milk and milk products with nanostructures is a new concept in the dairy sector. Various minerals and vitamins have been fortified by using various nanoencapsulation techniques. Minerals are less susceptible as compared to vitamins towards physical factors and chemical changes. Minerals are less stable when exposed to heat, air, or light. Fortification of minerals does not influence the sensory attributes or physicochemical properties of a dairy product. Nano peanut sprouts, nano-chitosan, nano-ginseng, and nano powdered eggshell have been used in dairy products. From safety point of view, there is a need to recognize possible interaction among NPs and biological systems. There is a need to study the *in-vivo* and *in-vitro* tests to find out the probability of toxicity in NPs in cells and in the living entities.

KEYWORDS

- absorption
- bioactive compounds
- dairy products
- emulsification
- food matrices
- nano encapsulation
- nanoparticles

REFERENCES

1. Abbasi, S., & Azari, S., (2011). Efficiency of novel iron microencapsulation techniques: Fortification of milk. *International Journal of Food Science & Technology, 46*(9), 1927–1933.
2. Abd-Rabou, N., Zaghloul, A., Seleet, F., & El-Hofi, M., (2010). Properties of edam cheese fortified by dietary zinc salts. *Journal of American Science, 6*(10), 441–446.
3. Ahn, Y. J., Ganesan, P., & Kwak, H. S., (2012). Comparison of nanopowdered and powdered peanut sprout-added yogurt on its physicochemical and sensory properties during storage. *Food Science of Animal Resources, 32*(5), 553–560.
4. Ahn, Y. J., Ganesan, P., & Kwak, H. S., (2012). Comparison of polyphenol content and antiradical scavenging activity in methanolic extract of nanopowdered and powdered peanut sprouts. *Journal of the Korean Society for Applied Biological Chemistry, 55*(6), 793–798.
5. Al Mijan, M., Lee, Y. K., & Kwak, H. S., (2014). Effects of nanopowdered eggshell on postmenopausal osteoporosis: A rat study. *Food Science and Biotechnology, 23*(5), 1667–1676.
6. Alzate, A., Pérez-Conde, M., Gutiérrez, A., & Cámara, C., (2010). Selenium-enriched fermented milk: A suitable dairy product to improve selenium intake in humans. *International Dairy Journal, 20*(11), 761–769.
7. Boccio, J., & Monteiro, J. B., (2004). Food fortification with iron and zinc: Pros and cons from a dietary and nutritional viewpoint. *Revista de Nutrição (Nutrition Journal), 17*, 71–78.
8. Bonnet, M., Cansell, M., Berkaoui, A., Ropers, M., Anton, M., & Leal-Calderon, F., (2009). Release rate profiles of magnesium from multiple W/O/W emulsions. *Food Hydrocolloids, 23*(1), 92–101.
9. Cayot, P., Guzun-Cojocaru, T., & Cayot, N., (2013). Iron fortification of milk and dairy products. In: *Handbook of Food Fortification and Health* (pp. 75–89). Cham: Springer.
10. Chang, Y. H., Lee, S. Y., & Kwak, H. S., (2016). Physicochemical and sensory properties of milk fortified with iron microcapsules prepared with water-in-oil-in-water emulsion during storage. *International Journal of Dairy Technology, 69*(3), 452–459.
11. Chauhan, A. K., Saini, R., & Kumar, P., (2019). Encapsulation of antioxidants using casein as carrier matrix. Chapter 10. In: Lohith, K. D. H., Goyal, M. R., & Suleria, H. A. R., (eds.), *Nanotechnology Applications in Dairy Science: Packaging, Processing, and Preservation* (pp. 245–258). Oakville - ON: Apple Academic Press.
12. Choi, K. H., Yoo, S. H., & Kwak, H. S., (2014). Comparison of the physicochemical and sensory properties of a Siago cheeses with added nano-powdered red ginseng and powdered red ginseng during ripening. *International Journal of Dairy Technology, 67*(3), 348–357.
13. Kumar, L. D. H., & Sarkar, P., (2018). Potential of nanotechnology in dairy processing: A review. Chapter 3. In: Goyal, M. R., (ed.), *Sustainable Biological Systems for Agriculture: Emerging Issues in Nanotechnology, Biofertilizers, Wastewater, Farm Machines* (pp. 55–80). Oakville - ON: Apple Academic Press.

14. Garg, S., Hemrom, A., Syed, I., Sivapratha, S., Panigrahi, S. S., Dhumal, C., & Sarkar, P., (2019). Nanoedible coatings for dairy food matrices. Chapter 2. In: Lohith, K. D. H., Goyal, M. R., & Suleria, H. A. R., (eds.), *Nanotechnology Applications in Dairy Science: Packaging, Processing, and Preservation* (pp. 27–42). Oakville - ON: Apple Academic Press.

15. Gaucheron, F., (2000). Iron fortification in dairy industry. *Trends in Food Science & Technology, 11*(11), 403–409.

16. Ghorani, B., & Tucker, N., (2015). Fundamentals of electrospinning as a novel delivery vehicle for bioactive compounds in food nanotechnology. *Food Hydrocolloids, 51,* 227–240.

17. Gulbas, S. Y., & Saldamli, I., (2005). The effect of selenium and zinc fortification on the quality of Turkish white cheese. *International Journal of Food Sciences and Nutrition, 56*(2), 141–146.

18. Gupta, C., Chawla, P., & Arora, S., (2015). Development and evaluation of iron microencapsules for milk fortification. *CyTA-Journal of Food, 13*(1), 116–123.

19. Gupta, C., Chawla, P., Arora, S., Tomar, S., & Singh, A., (2015). iron microencapsulation with blend of gum Arabic, maltodextrin and modified starch using modified solvent evaporation method-milk fortification. *Food Hydrocolloids, 43,* 622–628.

20. Gutiérrez, G., Matos, M., Barrero, P., Pando, D., Iglesias, O., & Pazos, C., (2016). Iron-entrapped niosomes and their potential application for yogurt fortification. *LWT, 74,* 550–556.

21. Hati, S., Makwana, M. R., & Mandal, S., (2019). Encapsulation of probiotics for enhancing the survival in gastrointestinal tract. Chapter 9. In: Lohith, K. D. H., Goyal, M. R., & Suleria, H. A. R., (eds.), *Nanotechnology Applications in Dairy Science: Packaging, Processing, and Preservation* (pp. 225–244). Oakville - ON: Apple Academic Press.

22. Huang, X., & Brazel, C. S., (2001). On the importance and mechanisms of burst release in matrix-controlled drug delivery systems. *Journal of Controlled Release, 73*(2, 3), 121–136.

23. Jafari, S. M., (2017). *Nanoencapsulation of Food Bioactive Ingredients: Principles and Applications* (p. 500). London: Academic Press.

24. Jain, E., Scott, K. M., Zustiak, S. P., & Sell, S. A., (2015). Fabrication of polyethylene glycol-based hydrogel microspheres through electro spraying. *Macromolecular Materials and Engineering, 300*(8), 823–835.

25. Kheadr, E. E., Vuillemard, J., & El-Deeb, S., (2003). Impact of liposome-encapsulated enzyme cocktails on cheddar cheese ripening. *Food Research International, 36*(3), 241–252.

26. Khosroyar, S., Akbarzade, A., Arjoman, M., Safekordi, A. A., & Mortazavi, S. A., (2012). Ferric-saccharate capsulation with alginate coating using the emulsification method. *African Journal of Microbiological Research, 6,* 2455–2461.

27. Kim, B., Kim, D., Cho, D., & Cho, S., (2003). Bactericidal effect of TiO_2 photocatalyst on selected food-borne pathogenic bacteria. *Chemosphere, 52*(1), 277–281.

28. Kim, D. H., Chang, Y. H., & Kwak, H. S., (2014). The preventive effects of nanopowdered peanut sprout-added Caciocavallo cheese on collagen-induced arthritic mice. *Korean Journal for Food Science of Animal Resources, 34*(1), 49–55.

29. Kumar, D. L., Mitra, J., & Roopa, S., (2020). Nanoencapsulation of food carotenoids. Chapter 7. In: Dasgupta, N., Ranjan, S. and Lichtfouse, E. (eds.), *Environmental Nanotechnology* (Vol. 3, pp. 203–242). Cham: Springer.

30. Kumar, D. L., & Sarkar, P., (2018). Encapsulation of bioactive compounds using nanoemulsions. *Environmental Chemistry Letters, 16*(1), 59–70.

31. Kumar, D. L., & Sarkar, P., (2017). Nanoemulsions for nutrient delivery in food. Chapter 4. In: Ranjan, S., Dasgupta, N., & Lichtfouse, E., (eds.), *Nanoscience in Food and Agriculture* (Vol. 5, pp. 81–121). Cham: Springer.

32. Kumar, L., & Sarkar, P., (2018). Potential of nanotechnology in dairy processing: A review. *Sustainable Biological Systems for Agriculture: Emerging Issues in Nanotechnology, Biofertilizers, Wastewater, and Farm Machines* (pp. 55–70). Oakville -ON: Apple Academic Press.

33. Kwak, H., Ju, Y., Ahn, H., Ahn, J., & Lee, S., (2003). Microencapsulated iron fortification and flavor development in Cheddar cheese. *Asian-Australasian Journal of Animal Sciences, 16*(8), 1205–1211.

34. Kwak, H., Yang, K., & Ahn, J., (2003). Microencapsulated iron for milk fortification. *Journal of Agricultural and Food Chemistry, 51*(26), 7770–7774.

35. Lee, S. B., Ganesan, P., & Kwak, H. S., (2013). Comparison of nanopowdered and powdered ginseng-added yogurt on its physicochemical and sensory properties during storage. *Korean Journal of Food Science, 33*(1), 27–30.

36. Maynard, A. D., Aitken, R. J., Butz, T., Colvin, V., Donaldson, K., Oberdörster, G., Philbert, M. A., et al., (2006). Safe handling of nanotechnology. *Nature, 444*(7117), 267–269.

37. Onsekizoglu, B. P., & Gunasekaran, S., (2017). Iron-encapsulated cold-set whey protein isolate gel powder-part 1: Optimization of preparation conditions and *in vitro* evaluation. *International Journal of Dairy Technology, 70*(1), 127–136.

38. Philippe, M., Le Graët, Y., & Gaucheron, F., (2005). The effects of different cations on the physicochemical characteristics of casein micelles. *Food Chemistry, 90*(4), 673–683.

39. Pirkul, T., Temiz, A., & Erdem, Y. K., (1997). Fortification of yoghurt with calcium salts and its effect on starter microorganisms and yoghurt quality. *International Dairy Journal, 7*(8, 9), 547–552.

40. Ramkumar, C., Vishwanatha, A., & Saini, R., (2019). Regulatory aspects of nanotechnology for food industry. Chapter 7. In: Lohith, K. D. H., Goyal, M. R., & Suleria, H. A. R., (eds.), *Nanotechnology Applications in Dairy Science: Packaging, Processing, and Preservation* (pp. 169–184). Oakville - ON: Apple Academic Press.

41. Rasenack, N., & Müller, B. W., (2004). Micron-size drug particles: Common and novel micronization techniques. *Pharmaceutical Development and Technology, 9* (1), 1–13.

42. Roblegg, E., Froehlich, E., Meindl, C., Teubl, B., Zaversky, M., & Zimmer, A., (2012). Evaluation of a physiological *in vitro* system to study the transport of nanoparticles through the buccal mucosa. *Nanotoxicology, 6*(4), 399–413.

43. Saeidy, S., Keramat, J., & Nasirpour, A., (2014). Physicochemical properties of calcium-fortified soymilk with microencapsulated and chelated calcium salt. *European Food Research and Technology, 238*(1), 105–112.

44. Sahoo, K. K., Tripathi, A. K., Pareek, A., Sopory, S. K., & Singla-Pareek, S. L., (2011). An improved protocol for efficient transformation and regeneration of diverse indica rice cultivars. *Plant Methods, 7*(1), 49–55.

45. Saini, R., Chauhan, A. K., & Kumar, P., (2019). Encapsulation of antioxidants using casein as carrier matrix. Chapter 1. In: Lohith, K. D. H., Goyal, M. R., & Suleria, H. A. R., (eds.), *Nanotechnology Applications in Dairy Science: Packaging, Processing* (pp. 245–257). Oakville - ON: Apple Academic Press.

46. Sathyashree, H., Ramachandra, C., Udaykumar, N. P., Naik, N., & DH, L. K., (2018). Rehydration properties of spray dried sweet orange juice. *Journal of Pharmacognosy and Phytochemistry, 7*(3), 120–124.

47. Seo, M., Lee, S., Chang, Y., & Kwak, H., (2009). Physicochemical, microbial, and sensory properties of yogurt supplemented with nanopowdered chitosan during storage. *Journal of Dairy Science, 92*(12), 5907–5916.

48. Seo, M. H., Chang, Y. H., Lee, S., & Kwak, H. S., (2011). The physicochemical and sensory properties of milk supplemented with ascorbic acid-soluble nano-chitosan during storage. *International Journal of Dairy Technology, 64*(1), 57–63.

49. Shivaram, S. H., & Saini, R., (2019). Spray drying-assisted fabrication of passive nanostructures: From milk protein. Chapter 3. In: Lohith, K. D. H., Goyal, M. R., & Suleria, H. A. R., (eds.), *Nanotechnology Applications in Dairy Science: Packaging, Processing* (pp. 45–68). Oakville - ON: Apple Academic Press.

50. Silvestre, M., Lagarda, M., Farré, R., Martınez-Costa, C., & Brines, J., (2000). Copper, iron and zinc determinations in human milk using FAAS with microwave digestion. *Food Chemistry, 68*(1), 95–99.

51. Singh, G., Arora, S., Sharma, G., Sindhu, J., Kansal, V., & Sangwan, R., (2007). Heat stability and calcium bioavailability of calcium-fortified milk. *LWT-Food Science and Technology, 40*(4), 625–631.

52. Spasova, M., Mincheva, R., Paneva, D., Manolova, N., & Rashkov, I., (2006). Perspectives on: Criteria for complex evaluation of the morphology and alignment of electrospun polymer nanofibers. *Journal of Bioactive and Compatible Polymers, 21*(5), 465–479.

53. Subash, R., & Elango, A., (2015). Microencapsulated iron for fortification in yoghurt. *Journal of Food Science, 6*(2), 258–262.

54. Ustunol, Z., & Hicks, C., (1990). Effect of calcium addition on yield of cheese manufactured with *Endothia parasitica* protease. *Journal of Dairy Science, 73*(1), 17–25.

55. Wagner, M. E., Spoth, K. A., Kourkoutis, L. F., & Rizvi, S. S., (2016). Stability of niosomes with encapsulated vitamin D3 and ferrous sulfate generated using w novel supercritical carbon dioxide method. *Journal of Liposome Research, 26*(4), 261–268.

56. Woestyne, M. V., Bruyneel, B., Mergeay, M., & Verstraete, W., (1991). The Fe^{2+} chelator proferrorosamine A is essential for the siderophore-mediated uptake of iron by *Pseudomonas roseus fluorescens*. *Applied Environmental Microbiology, 57*(4), 949–954.

CHAPTER 5

NANOPARTICLES AND NANO-FORMULATIONS FOR FOOD APPLICATIONS

NAINSI SAXENA, S. S. SHIRKOLE, SAHELY SAHA, and
B. MANJULA

ABSTRACT

The application of nanotechnology in food has garnered significant scholarly attention for its potential in the food industries. The functional properties of nano-formulations and nanoparticles (NPs) with high efficacy at low levels can be used in different food applications. These NPs are promising candidates for food packaging and nutrition. However, the toxicity of the NPs is still questionable and whether they may be mass-produced in near future. This chapter overviews different NPs and their functional properties associated with food applications.

5.1 INTRODUCTION

Nanomaterials are elucidated with size of 1 to 100 nm and possess not less than one of the three dimensions [10, 52]. For the reason that, nanoparticles (NPs) acquire high surface area per unit volume. Accommodation of high fractions of atoms near the surface layers and in the surface demonstrates the ability of NPs to exhibit quantum effects. However, the unique functional properties of NPs cannot be predicted by extrapolation of bulk material properties [1].

NPs exist in the form of metal oxides, metals, polymers, carbon materials and biological structures [1]. The morphological diversity of

NPs (cylinders, platelets, disks, hollow spheres, tubes, etc.), demonstrates the possible ways to control their properties [22]. Currently, synthetic routes (such as liquid, gas, and solid phase) are used to prepare the NPs. However, NPs are chemically reactive and in many cases surface functionalization such as chemical grafting are required to stabilize [4, 51, 76]. Nevertheless, these surface functionalization techniques are specific based on requirement of specific application [45]. Hence, NPs are essential building blocks for several nanotechnology applications.

NPs are gaining importance in food processing with improved properties due to their characteristic size and dimension. Such properties might include strength, thermal conductivity, and elasticity [2, 76]. NPs can nonspecifically bind to proteins, enzymes, antibodies, or nucleotides, resulting in modulation of the cell signaling [12].

For medicinal applications, NPs must conform to a size smaller than 100 nm, and have uniform physicochemical properties and be nontoxic and biocompatible [36, 76]. Research studies on nanotechnology for formulation of NPs reinforced composites, delivering of functional molecule in fried foods is creating new trends in sustainable food processing [9]. However, in order to increase the nanotechnology applications in food, novel engineered nanostructures must be formulated and their special organization should be known (Table 5.1). Nevertheless, their toxicity is a major concern due to unmeasured risks caused by bioaccumulation.

Nanotechnology field is currently revolutionizing food industry due to nanoencapsulation of functional compounds. Major sources for functional compounds are natural products, which are therapeutically most important. Identification and discovery of novel functional compounds and their respective properties for functional food development has always been in demand. Encapsulation of such functional phytochemicals using suitable delivery matrix can enhance their bioavailability in gastro-intestinal tract by altering their biodistribution and pharmacokinetics [43]. In addition, therapeutic index of functional compounds in the encapsulation matrix has been shown to increase due to controlled delivery and specific localization [43, 44, 47, 66].

Recent trends in food-pharma industry have shown that nano-based functional compounds have potential of being commercialized and there is an upsurge of nano-products. Nevertheless, these days, consumers also attracted towards natural dietary agents due to their proven health benefits. For instance, "taurine" (a functional compound extracted from cactus pear) has the potential nutraceutical property, such as, antidiabetic, antiviral, and anticancer. Due to cumulative interest in traditional medicine, World Health Organization (WHO) has introduced a global approach for enhancing knowledge of traditional medicine. Recent studies on nano-formulations of functional compounds have been to eliminate the limitations associated with them and maximize their health benefits.

This chapter focuses on different nano-formulation strategies associated with different functional compounds. This chapter also reviews the different NPs properties and their applications in food industry.

5.2 NANO-NUTRACEUTICALS AND NANO-THERAPEUTICS

Nutraceuticals can be a part of food or food itself, which offers pharmaceutical and nutritional benefits. However, functional foods consist of different health promoting compounds, such as: vitamins, protein-peptides, bioactive lipids, probiotics, and minerals. Hence, nutraceuticals can also be considered as elements of food, which can enhance body functions by fighting against determined diseases. The recent advancements in the field of food biophysics have opened the gate for the understanding of food matrices, which can help in studying incorporated functional molecule fate during storage and digestion. Efficacy of any functional compound is reliant on its bioavailability [32].

In nutritional terms, bioavailability refers to the percent availability of a compound at the site of action. With increased health concerns and awareness of nutraceuticals, their bioavailability has become a challenge to the manufactures and regulators of such products. For instance, when a functional compound needs to administered orally, different constraints, such as, solubility in guts, gastric residence time, permeability, stability under different food processing conditions and gastrointestinal tract decreases the functional properties of various phytochemicals. Along with their low bioavailability, crystallization, and chemical stability also need to be addressed before incorporation into the food matrix. Hence, application of nanotechnology in nutraceuticals can overcome these challenges and help in developing nano-enabled functional foods [8, 19].

Nanotherapeutics is application of nanotechnological principles that has wide range of impact on humans. The nanotherapeutics branch is exploiting a wide range of nanotechnological approaches including numerous biosensors and biological devices. The quantum reactions taking place at nano-scale have an impact on physicochemical, biological, optical, and mechanical characteristics, which permit researchers to explore the benefits of such phenomena. Nanotherapeutics also comprises novel emerging concepts for designing of nano-machines called nano-robots. The unique property of nanomaterials is their size, which is similar to many biological macro-molecules facilitating the use nano-bioactive compounds in *in-vitro* and *in-vivo*. Thus, amalgamation of nanotechnology with biological molecules, physiotherapy, analytical tool, and drug delivery vehicles can enhance the potential of technology. Hence, in the light of evidences mentioned here, nanotherapeutics has significant applications in healthcare management.

In general, nanotechnology strategies are meant to enhance the bioactive compound therapeutic concentration, absorption, and stability resulting in higher effective targeted delivery. In addition, nanotechnology has been utilized in controlled release of bioactive compounds in digestive tract. For instance, nanoelectromechanical system has potentially essential role in treatment of cancer using gold and iron NPs. In this technique, overdose of drugs or functional compound can be avoided and level of therapeutic compound can be monitored in the body. Rational design of nanotherapeutics involves the formulation of nano-templates/platforms of a particular size, shape, charge, and other surface properties that are essential for interaction with biological molecules resulting in therapeutic effect [21, 61]. Table 5.1 indicates impact of NPs in food industry.

TABLE 5.1 Engineered Nanoparticles and Nanotechnology Impact in Food Industry: Present and Potential

Nanotechnology	Present Impact	Potential Impact
Coating	Edible coatings	Active packaging materials, multifunctional coatings
Dispersion	Delivery systems, absorption, or desorption materials	Targeted delivery systems
High surface area materials	Encapsulation, catalysis	Designer molecules for the protection of functional compounds
Nano devices	Sensors	Multifunctional sensors for the detection of wide range of microorganisms of same trait
Nanocomposites	Packaging materials	Nanoparticles-filled polymer composites

5.3   PREPARATION OF NANOPARTICLES (NPS)

Nanotechnology comprises of mainly of production, manipulation, and use of materials in the nano range from less than a micron to that individual atom [63]. There are two NPs preparation approaches broadly classified for food application, such as: "top-down" or "bottom-up" [46, 81]:

1. **The Top-Down Approach:** It is a process via reduction in size of particles using physical processing, such as: homogenization or milling. For instance, different food matrices (such as: wheat flour, green tea powder) can be produced by dry milling. When the larger particles are reduced into smaller size particles in some nano meter range with increase in surface area, then such a material will show enhancement in its properties.

2. **The Bottom-Up Approach:** It utilizes self-assembly property of the individual particle. Self-assembly can occur in wide range from nanometer scale to larger size, which include various covalent and non-covalent interactions. For example, nanoscale structures can be achieved via balance of contrasting non-covalent interactions, e.g.: globular proteins and organization of casein micelle.

Many other methods come under these techniques, such as: chemical synthesis and biosynthesis. Chemically synthesized NPs demonstrate toxic effect on human health and environment. Biologically synthesized NPs mainly utilize microorganisms, fungi, algae, and enzymes [73]. Recently green synthesis of NPs method also been introduced [5, 70]. These eco-friendly methods are useful in different areas for food industry especially metallic NPs, such as: Ag, TiO_2 and Zn NPs.

5.4   CLASSIFICATION OF NANOPARTICLES (NPS) AND THEIR FUNCTIONAL PROPERTIES

5.4.1   INORGANIC NANOPARTICLES (NPS)

Inorganic NPs properties (such as: chemical inertness, good stability, and ease of functionalization) are critical in selection of NPs for different applications [7]. The stability of NPs and optimized functional properties make inorganic NPs more applicable than the organic ones. All the metal

NPs (such as: TiO_2, Zinc, and AgNPs) come under this category, which have wide applications in food sector, such as: food packaging, sensing, and for antimicrobial activity [25, 54]. Biodegradability of packaging materials can enhance mixing with the inorganic NPs, such as: clay is added into the bio-polymeric matrix; and AgNPs are added for keeping the food fresher for long time as antimicrobial agent in food packaging [72].

5.4.2   ORGANIC NANOPARTICLES (NPS)

Most of the natural NPs come under organic NPs (such as: glucose, proteins, carbohydrates, lipids, polysaccharides (chitosan-based nanomaterials), casein (micelle presented in milk)) that are frequently used in food industries. The small size and shape with unique properties keep organic NPs always at first place by different research communities for study and application [75].

5.4.3   NATURAL NANOPARTICLES (NPS)

Some important types of natural NPs and their size ranges which are useful in food sector are mentioned in Table 5.2. In a specific focus on food industry, NPs can be classified mainly three types, such as: natural, incidental, and engineered NPs:

- Natural nanoparticles are those nanoparticles having one or more dimension and generated by various natural processes and in particular environment conditions, such as: lipids, proteins (casein micelle in milk), carbon nanoparticles caramelized foods, foam, magneto-tactic bacteria, and chitosan-based nanoparticles [87].
- Incidental nanoparticles are the result from either manmade or industrial processes, such as: milling, grinding, and combustion.
- Engineered nanoparticles are those nanoparticles, which are manufactured by modifying the specific properties and composition [17].

5.5   DIFFERENT NANO-FORMULATIONS

An efficient formulation matrix should encapsulate and protect the functional compound for the successful therapeutic activity. It is possible to design a targeted formulation matrix that releases the bioactive compound to a specific site of action (such as: mouth, stomach, small intestine, large intestine, or colon) and to protect of the bioactive compound against chemical and enzymatic degradation and potentially enhance the selective uptake of these formulations. In recent years, numerous nano-formulations have been developed to improve the bioactive delivery system, such as: polymeric NPs, nanogels, nanoemulsions, nanostructured lipid carriers, solid lipid NPs, dendrimers, nano-capsules, and nano-sponges.

TABLE 5.2 List of Important Natural Nanoparticles in the Food Sector and Their Size Ranges

Natural Nanoparticles	Size range (nm) (Length or Diameter)
Casein micelle	60–100
DNA	12
Glucose	21–75
Liposome	30–10,000

Ideally, synthetic polymers are used to fabricate polymer NPs. The natural polymers due to variation in constancy and purity from batch-to-batch results in poor controlled release profile and reproducibility of entrapped bioactive compound. Lipophobic moieties can be encapsulated into polymer NPs stabilized double emulsion matrices, because it is difficult to maintain the functionality of functional molecule in presence of organic volatile solvents. Hence, polymeric NPs can protect unstable phytochemicals from degradation, thereby reducing the side effects of toxic drugs. In addition, incorporation of health benefiting molecules in lipophilic cavity of NPs can ensure the desired *in-vivo* and *in-vitro* effects.

Interaction of the polymeric matrix with different environmental parameters (such as: temperature, enzymes, pH, and ionic strength) causes degradation. This allows to control the concentration of bioactive compounds that are released over time. The bioactive component release profile and rate depends on polymeric degradation rates and changes in the diffusion process, the thickness of the encapsulating shell and the material used in the coating. A longer release time lasted for thicker shells [9].

Examples for polymeric NPs-based formulations available in market are: Abraxane, Enantone Depot®, Gonapeptyl Depot®, Decapeptyl®. Nano-capsules are another type of nano-formulations matrices, where these comprise of either solid or lipid core, in which bioactive compound is loaded and encapsulated by polymer matrices. Nano-capsules can also produce by a complex coacervation method using two solutions of oppositely charged polymers and their electrostatic interactions.

In a recent study, resveratrol (Rsv) encapsulated lipid core nano-capsules were formulated for targeting colon cancer cells. The results demonstrated that nano-capsules enhanced the anticancer effect and exhibited controlled and sustained release of bioactive compound. Nanogels, consisting of flexible lipophobic polymer matrix, can be fabricated as plain gels. The functional compound can be encapsulated spontaneously in the nanogel by diffusion on swelling solvent/water. Due to high diffusion, gel matrix collapses and results in formation of dense solid nanoparticle with decrease in solvent concentration.

Nanogel matrices possess different unique properties, such as: high moisture, biocompatibility, and desirable mechanical strength among nano-formulations. Different approaches (such as: micro-molding, photo-lithographic, continuous micro-fluidics, heterogenous polymerization and biopolymer modification for fabrication of nanogels) have been investigated. Several criteria are considered for the formulation of nanogels, such as: stability of nanogels in gastrointestinal tract, site specificity, ionic strength, pH, temperature, and biodegradability without compromising the encapsulated functional compound efficacy. Few examples for nanogels at commercial scale include: Oxalgin and Sane-Care Nanogel.

Nanoemulsions can be classified as oil-in-water (milk) and water-in-oil (butter). In general, nanoemulsions are thermodynamically stable colloidal dispersions, translucent, and interface of two liquids that are stabilized by an amphiphilic molecule. Nanoemulsions help to overcome unpleasant taste, prevents against oxidative and hydrolysis of bioactive lipids. In recent years, nanoemulsions are utilized in formulation of sanitizers, flavored beverages and nutraceutical dairy and dessert foods. Commercially available nanoemulsion based formulations are Troypofol, Ropion, Liple, and Norvir.

Lipid NPs fabricated with solid matrices are called solid lipid NPs. They are synthesized using oil-in-water nanoemulsion with a solid lipid. The benefits of solid lipid NPs are: low raw material cost, ease of scale-up, enhanced bioavailability and biocompatibility, less/no use of organic solvents and controlled release of bioactive compound. However, expulsion phenomena, polymorphic transition, uncontrolled gelation, high crystallinity of solid lipids and less loading capacity make them less suitable for bioactive compound delivery. On the other hand, nanostructured lipid carriers comprise of two-level nano-lipids, i.e., solid lipid NPs are encapsulated into liquid lipids. These nanoformulations help in prevention of coalescence of particles and permit immobilization of therapeutic compounds. In addition, due to presence of oil dispersions in solid matrix, their encapsulation efficiency is high in comparison to solid lipid NPs.

Nanostructured lipid carriers have been investigated for the delivery of lipophilic and lipophobic bioactive compounds. Due to low toxicity, no organic solvents, controlled release, and biodegradability properties of nanostructured lipid carriers make them superior to polymeric NPs [18, 22, 23, 27]. Dendrimers are three-dimensional branched polymeric structures with high water solubility, narrow poly-dispersibility, nano-size, and tailorable molecular structure. The terminal end of dendrimers functions as a template/platform for the site targeting and conjugation. In addition, peripheral functional groups enable their tailor-made features, enhancing their versatility. Besides their enhanced solubility, dendrimers have been explored as formulation matrices for oral, pulmonary, and ocular routes.

5.6   FUNCTIONAL PROPERTIES OF NANOPARTICLES (NPS)

The applicability and importance of NPs depends on particle agglomeration, surface charge, size distribution, particle size and shape [71]. The most important concerns of the nanoscale structures are the existence of a high fraction of molecules/atoms instituting the nanoparticle on the nanoparticle surface instead of in the interior surface of a nanoparticle and the highly available surface area per unit volume also influences the toxicity and functionality of NPs [29]. However, the magnitude of fraction of molecules/atoms and surface area to volume ratio are increased with decrease in particle size. The distinctive functional biological, chemical, and physical properties of NPs initiate from these two features [48, 55]. In addition, NPs quantum effects are demonstrated consenting for a numerous application in different areas [67]. Further, unusual structural morphologies of some nanostructures (such as: dendrimers and carbon nanotubes (CNTs)) contribute to structural-morphology dependent innovative applications.

The optical properties (such as: emission and absorption of wavelength) can be controlled by surface functionalization and particle size [40].

The particle size and chemical nature of nanoparticle size regulate the electron affinity or ionic potential and thus these define the electron transporting properties of NPs [14, 62]. However, reducing particle size of a metal nanoparticle can decrease the melting and sintering temperature. Surface functionalized NPs can be integrated into the solid matrix to offer better thermal conductivity. In addition, decrease in particle size can enhance the magnetic behavior of some metal oxides and metals; whereas, distinct metallic NPs possessing magnetic properties can demonstrate superparamagnetic behavior [57, 58, 85].

The catalysis is improved by homogenous distribution and large surface area to volume ratio of NPs [13]. High surface to volume ratio provides strong interactions among the matrix and NPs; nevertheless, in composites it depends on aspect ratio, chemical nature, interactions at interface (controlled by surface modification) and extent of dispersion [2, 20].

In addition, different unique properties of composites can be obtained through surface functionalization of NPs. In particular, it is possible to obtain NPs composite with high elastic modulus; wherein, less impact strength is observed compared to composite formulated with larger particles [26]. The barrier properties of silicate NPs-based polymer membranes are enhanced by large specific surface area and platelet morphology, which can be used for enhancing the pathways of molecular transport of infusing materials. In addition, NPs can also stimulate the flammability of polymers by increasing the heat deflection temperature and glass transition temperature [41, 66, 74].

5.7   NANOENCAPSULATION OF POLYPHENOLS AND ANTIOXIDANTS

Nanoformulations may facilitate the supplementation of nutrients and their bioavailability to uplift the health of individuals. The nanoparticle can stabilize numerous phytochemicals and enhance their cellular absorption. Due to their small size and high surface area to volume ratio, they can transport faster in digestive tract and enhance the delivery of nutrients. In addition, different advantages of considering encapsulation in food industry are: retention of volatile molecules, taste masking, environmental stress (pH, ionic strength, moisture) triggered release, multiple bioactive compounds delivery using co-encapsulation technique, protection against oxidation, reducing reactions between food matrix micro/macro-molecules and bioactive compounds and formulation of optically transparent beverages fortified with phytochemicals [53].

It has been widely acknowledged that the chemical structure of polyphenols is significant to determine the rate and extent of their absorption upon ingestion rather than their concentration. The cell-membrane permeability and aqueous solubility of polyphenols are other factors that influence their bioavailability. For instance, curcumin is less soluble and possesses low cell membrane permeability. Epigallocatechin gallate is highly soluble with low cell membrane permeability. On other hand, Rsv is a low soluble polyphenol with high cell permeability.

The main factors affecting polyphenol bioavailability are low absorption in the gastrointestinal tract, extensive transformation within the gut, rapid metabolism, and systemic elimination. In particular, polyphenols interact with proline-rich salivary proteins forming insoluble complexes. They are exposed to the stomach acid conditions and the alkaline status of the small intestine, which are deleterious for their stability; polyphenols are subjected to extensive first-pass phase II metabolism, principally methylation, sulfation, and glucuronidation, in the small intestine; and then they are metabolized in the liver. These metabolic transformations cause significant changes in polyphenol structure and biological activity.

Consequently, the forms able to reach the blood and tissues are different from those present in food. Moreover, it is challenging to identify all polyphenol metabolites and evaluate their biological activity. Different encapsulation materials at nanoscale are used to entrap the polyphenols. For instance, entrapping of EGCG in beta-lactoglobulin NPs showed higher binding efficiency and stability compared to the native beta-lactoglobulin EGCG complex.

5.8   EFFECT OF FOOD MATRIX PROPERTIES AND COMPOSITION ON STABILITY AND RELEASE OF BIOACTIVE COMPOUNDS DURING DIGESTION

Considering the constraints of encapsulation technique in real food matrix, different food matrix properties (such as: porosity, compositional variation, presence of surface-active compounds), minerals, and other hydrocolloids can intervene in the stability of a bioactive compound. In addition, different digestion stages involve different environmental conditions, which differ from each other, such as: pH, enzyme composition, nutrient absorption and/or adsorption, mixing space and residence time. Dietary fats can increase polyphenol bioavailability by increasing absorption, possibly by enhancing micellization in the small intestine.

Studies have demonstrated that polyphenolic extracts from different foods can exhibit different bioaccessibility in combination compared to the individual effect. Green et al. studied the effect of matrix composition (bovine, rice milk, soya milk, citric acid) on enhanced bioaccessibility of green tea catechins. They concluded that matrix ingredients help in protection and stabilization of bioactive compound from auto-oxidation in alkaline condition [28].

The bioaccessibility of epigallocatechin and epigallocatechin gallate in green tea was increased after association with ascorbic acid and sucrose simultaneously, which was attributed to enhanced uptake of Caco-2 cells and bioavailability in rats [59]. In another study of co-digestion with milk and blueberries, the recovery of total phenols and anthocyanins was reduced due to milk [11]. Studies with raspberry juice demonstrated the influence of ice-cream physical properties on anthocyanins recovery. In addition, co-digestion of the raspberry extract with combined foodstuffs (such as: bread, breakfast cereal, ice cream, and cooked minced beef) gave a different pattern [49]. The reduced recovery of anthocyanins can be due to pH and they can generate insoluble complex influencing the absorption rate [49].

During the digestion process, bioactive compounds release in different gastric phase. The transition from acidic gastric to the mild alkaline intestinal environment influences the stability and absorption rate of phytochemicals. The overall bioavailability process includes gastrointestinal digestion, absorption, and metabolism. Thus, in the gastrointestinal tract, anthocyanins may be released from the food matrix, modified under the influence of digestive enzymes as a result of pH changes. The gastric phase is usually the location, where food components are mostly dissolved due to action of pH, peristaltic movement, and pepsin digestion [50].

The low pH gastric phase helps the bioactive compounds transition from the matrix into aqueous phase. The enzymes secreted in small intestine involves in digestion of different categories of food components (such as: lipids, phytochemicals, micronutrients) and forms the water-soluble mixed micelles. The digestive stability of different food components can be assessed by *in vitro* digestion method. For instance, the gastric treatment of raw red cabbage enhanced significantly anthocyanins content (by 62.7%), but total phenolics content was increased only by 25.1%. On the other hand, after a gastric digestion of anthocyanin-rich extract increase of 16.9% in the total phenolics content was apparent. Differences between the results for red cabbage and isolated anthocyanin-rich extract suggest that concentration of anthocyanins released during digestion process and their stability strongly depend on the food matrix [60].

5.9   TOXICITY OF NANOPARTICLES (NPS)

In general, common NPs expected to be present in food matrix and food packaging are TiO_2, SiO_2 and metallic silver [82]. However, many food structures already contain natural NPs; while, concern is on the "manufactured" or "engineered" NPs and their effects [6]. In addition, NPs can migrate into foods in case of nanoparticle-based packaging materials. NPs consumed through food can enter into the cell matrices, where macro and micro particles cannot. On the other hand, it is less evident that particle size alone is responsible for this effect [34, 86].

Soluble nanostructures (such as: gels, emulsions, or colloids) engineered to deliver the bioactive compounds into foods and any effect caused by these nanostructures by interactions with body fluids or food acids can also cause toxicity in body via bioaccumulation [86]. For instance, colloidal delivery systems can carry titanium, silicon, and silver into body; while, in case of silver, it is not bioactive until the silver metal is transformed to silver ions and it can become harmful to body [3].

NPs can cause severe or acute lethal health effects. The toxic nature of NPs depends on: (i) inflammatory effects per unit mass; (ii) method of formulation [33]. In brief, higher surface area per unit volume of NPs potentially increases its toxic nature; while, NPs engendered during any industrial process can also exhibit adverse health. Under such conditions, NPs surface chemistry play a vital role in addition to size of particles. For instance, polytetrafluoroethylene (PTFE) is toxic to humans, mammals, and birds. The fumes generated from PTFE comprise of nanostructures (18 nm), which exhibit high toxicity and can cause polymer fume fever [39].

Unlike the toxicity of bulk particles, the toxic effects of NPs are modulated by their size, shape, topology, surface functionalization and aggregation status [39, 80]. Toxicity of one-dimensional and two-dimensional

NPs also depends on the particles aspect ratio (length/diameter) [24]. The NPs unique properties dictate their fundamental characteristics, including their ability to get into cells and cause toxicity. The following factors may determine their toxicity:

- For particle diameter less than 100 nm, a decrease in size increases the particles surface area and reactivity and their potency to cause adverse effects [30, 79].
- Carbon nanoparticles are unique because, depending on their folding patterns, they may exhibit different properties. Aggregation state of nanoparticles is affected by interparticle Vander Wall interactions [56]. The aggregated nanoparticles are much larger than the primary size of the particles. The properties and mobility of aggregated and dispersed nanoparticles are considerably different in the host environment. The aggregation behavior of nanoparticles and their adverse effects depend on their functionalization, hydrophobicity, length, and diameter. Aggregated nanoparticles mostly accumulate in cytosol and lysosomes. This suggests that aggregation compromises the relation between the nanoparticle size and toxicity, thus compromising their risk assessment [37, 84]. As discussed, size is an important issue when determining the nanoparticle dose in animal experiments.
- The dose in terms of mass density (such as: g per kg) may not be appropriate for nanoparticles. Other metrics, such as surface area or surface activity per unit body weight may be more relevant [69].
- Functionalization of nanoparticles such as enhancing or hindering of hydrophobicity explains the modification in metabolism of nanoparticles [15]. The hydrophobic nanoparticles agglomerate is less bioavailable in aqueous medium; however, chemical functionalization prevents the hydrophobic aggregation and improves the bioavailability [37].
- In addition, surface functionalization also improves the targeted functionalities and site-specific release of functional molecules. Nevertheless, dispersion media play an essential role in toxicity exposure of nanoparticles.
- Aggregated nanoparticles lack the unique properties of dispersed particles. Thus, synthetic dispersants and surfactants, which bind to the nanoparticles non-covalently, are commonly used. Such noncovalent functionalization has great promise, because the procedure

minimizes alterations in the electronic and mechanical properties of nanoparticles.

- However, it is important to ensure that the dispersants alone or the nanoparticle-dispersant complex are nontoxic. Both exposure and hazard (toxicity) are critical in nanoparticles risk assessment. For instance, though the nanoparticle is highly toxic, may demonstrate low risk, if exposure is very low. Nevertheless, there have been few studies which considered both exposure and toxicity together [78].
- Conventionally nanoparticles exposure monitoring depends on characterization of mass and bulk chemistry of nanostructured particles.
- Skin penetration, inhalation, and ingestion are potential exposure routes for engineered nanoparticles.
- For the evaluation of exposure of nanoparticles factors such as shape, surface chemistry, surface area, particle size and charge are considered [64].

5.10   RISK ASSESSMENTS OF NANOPARTICLES (NPS)

In food matrices, NPs are present at levels of parts per million (ppm) and they often are mixed with bulk particles to define the structural morphology and functionality. For instance, food matrices often contain anti-clumping agent silica (silicon dioxide) to keep food matrix free-flowing; while, titanium dioxide for enhancing the whiteness (frosting of doughnuts) [83]. Nevertheless, antimicrobial silver nanoparticle is restricted in food; on the other side, residues may present on vegetables and fruits that have been cleaned by washing with nano-silver suspension [16]. Risk assessment is difficult due to lack of NPs toxicological studies and morphological changes in digestive system after ingestion [68]. However, the risk associated with NPs is different to those with bulk particles and these properties should be considered by research groups, industry, and government regulatory agencies during regulatory steps and toxicology consideration [38, 65].

In addition, homogenous NPs may show the similar properties to that of bulk materials, but the toxic issues may not be same in either case [31, 42, 77]. For the regulatory purpose, NPs are considered as a variation of technical particles current formulation; and accordingly, it does not require distinct legislation [35]. In the next 10 years, the use of nanotechnology and NPs is expected to dramatically increase in food industry. In fact, currently NPs or nanomaterials (such as: proteins, composites, and emulsions) are being used in food industry on a large scale. Although there is unawareness about these technological developments among the public, yet it would be prudent in addressing human health and environmental concerns before a widespread utilization of NPs in food.

5.11 CONSTRAINTS AND FUTURE SCOPE OF NANO-FORMULATIONS

Nano-formulations exhibit unique challenges as there is not enough research data available on characterization of safety data, classification, analysis, and accumulation at the cellular level. Nanostructures show significant variation in their biological activity and toxicity based on surface chemistry. In addition, nanostructures demonstrate a huge health threat since they are capable of penetrating the barriers present within the body and reaching the biological systems. There are research studies available evidencing the effect of nanomaterials on cellular organelles and membranes owing to generation of free radicles.

Nanomaterials either present intracellularly or adsorbed to the cell surface can stimulate an immune response by reacting with a receptor present on the cell surface. Nanotoxicity assessment of new nanostructures should be given due importance in early stage of production. The in-depth understanding of absorption-distribution-metabolism-excretion nanomaterials and the effect of critical products features on them would help to develop model techniques with the view to anticipating the biological and toxic effect of nano-active products.

The substrate rigorousness and nano-topographical studies have been employed to study the effect of NPs on cell behavior, such as: adhesion, spreading, differentiation, and proliferation. The alteration of cellular surface may give rise to variation in physiological, developmental, and pathological processes.

Considering the process limitations, nanotechnology needs more scale-up studies and projection of different possible solutions during food storage and processing. For instance, liposomal formulations are recommended to be stored at refrigeration temperature, but not in a freezer to evade disruption of lipid bi-layers. In developing these matrices, it is essential to consider the end-application, the limitations and benefits offered by encapsulation over the presentation of a neat bioactive. The effectiveness of nano-encapsulated compounds must also be tested to ensure that the process is actually improving the characteristic of non-encapsulated compounds. On the other hand, each nano-formulations should be considered unique and evaluated individually.

In addition, cutting edge techniques such as surface engineering to tailor a delivery system is required to optimize carrier performances. Hence, studies on nano-formulation matrices surface modification and controlling based on encapsulation of desired bioactive compound will lead to achieve a high encapsulation efficiency and controlled cellular uptake for extended period of action.

Following conclusions based on literature review in this chapter can be mentioned:

- Evolution of nanotechnology has changed dramatically over the last decade and a significant focus on nanomaterials and application of nanotechnology has occurred in many areas of science. Nanoscale materials have, of course, always existed, yet remained largely undiscovered by science until timely scientific breakthroughs allowed us to resolve them. In recent years utilization of nanoparticles in food introduced to new era. However, their toxicity is the major concern due to unmeasured risk caused by bioaccumulation.

- The limitations in our knowledge are partly due to the lack of methodology for the detection and characterization of engineered nanoparticles in food like complex matrices. It is clear from this chapter that the field of nanotechnology has stemmed from the original work of only a handful of scientists, who postulated the possibilities of nanotechnology long before widespread understanding of the field.

- In the concepts of bioavailability and solubilization, it is essential to control, trigger, and sustained release of bioactive compounds and nutritional additives into food matrix as well as body.

- It is essential to formulating a nano-delivery vehicle with certain properties or to implant within them that which will control the release of bioactive compound from food matrix into our body. However, pharmaceutical industry has progressed in the field of controlled release of functional compounds, but these are at high cost.

- In addition, in food industry, the stability of entrapped bioactive compound during processing and storage is the first priority rather than controlled release. Hence, considering the need of evaluating different nano-formulations for their toxicity, it is crucial to assess them for their end-use application in different processing sectors.

5.12   SUMMARY

The introductory section explores why the behavior of nanoscale materials may differ from that of bulk and their atomic counterparts, and how it impacts the food structure to rationalize this point. Functional properties and functionalization of NPs are described in more details, although an attempt was made to keep the level of discussion quite simple. In last, the concept of NPs toxicity and its assessment is explained. This topic will be of fundamental importance for future applications in food industry. The nano-formulations aimed to discuss different nano-formulations for various hydrophilic and lipophilic bioactive compounds. Authors have discussed need of nano-formulations followed by describing nano-therapeutics and nano-nutraceuticals. Later sections have briefed on various nanoencapsulation matrices and functional compounds encapsulation in different encapsulation systems.

KEYWORDS

- absorption
- anti-clumping agent
- antimicrobial activity
- antimicrobial agent
- aqueous medium
- bioaccumulation
- bioavailability
- biocompatible
- biodegradability

REFERENCES

1. Aider, M., (2010). Chitosan application for active bio-based films production and potential in the food industry: Review. *LWT-Food Science and Technology, 43*(6), 837–842.

2. Arora, A., & Padua, G., (2010). Review: Nanocomposites in food packaging. *Journal of Food Science, 75*(1), R43–R49.
3. Ayala-Núñez, N. V., Villegas, H. H. L., Turrent, L. D. C. I., & Padilla, C. R., (2009). Silver nanoparticles toxicity and bactericidal effect against methicillin-resistant *Staphylococcus aureus*: Nanoscale does matter. *Nanobiotechnology, 5*(1–4), 2–9.
4. Balazs, A. C., Emrick, T., & Russell, T. P., (2006). Nanoparticle polymer composites: Where two small worlds meet. *Science, 314*(5802), 1107–1110.
5. Bar, H., Bhui, D. K., Sahoo, G. P., Sarkar, P., De, S. P., & Misra, A., (2009). Green synthesis of silver nanoparticles using latex of *Jatropha curcas*. *Colloids and Surfaces A: Physicochemical and Engineering Aspects, 339*(1), 134–139.
6. Baun, A., Hartmann, N. B., Grieger, K., & Kusk, K. O., (2008). Ecotoxicity of engineered nanoparticles to aquatic invertebrates: Brief review and recommendations for future toxicity testing. *Ecotoxicology, 17*(5), 387–395.
7. Birnbaum, D. T., Kosmala, J. D., & Brannon-Peppas, L., (2000). Optimization of preparation techniques for poly (lactic acid-co-glycolic acid) nanoparticles. *Journal of Nanoparticle Research, 2*(2), 173–181.
8. Birwal, P., Rangi, P., & Ravindra, M. R., (2019). Nanotechnology applications in packaging of dairy and meat products. Chapter 1. In: Lohith, K. D. H., Goyal, M. R., & Suleria, H. A. R., (eds.), *Nanotechnology Applications in Dairy Science: Packaging, Processing, Preservation* (pp. 3–26). Oakville - ON: Apple Academic Press.
9. Borel, T., & Sabliov, C., (2014). Nanodelivery of bioactive components for food applications: Types of delivery systems, properties, and their effect on ADME profiles and toxicity of nanoparticles. *Annual Review of Food Science and Technology, 5*, 197–213.
10. Calzolai, L., Gilliland, D., & Rossi, F., (2012). measuring nanoparticles size distribution in food and consumer products: A review. *Food Additives & Contaminants: Part A, 29*(8), 1183–1193.
11. Cebeci, F., & Şahin-Yeşilçubuk, N., (2014). The matrix effect of blueberry, oat meal and milk on polyphenols, antioxidant activity and potential bioavailability. *International Journal of Food Sciences and Nutrition, 65*(1), 69–78.
12. Chen, L., Remondetto, G. E., & Subirade, M., (2006). Food protein-based materials as nutraceutical delivery systems. *Trends in Food Science & Technology, 17*(5), 272–283.
13. Chen, L. Q., Fang, L., Ling, J., Ding, C. Z., Kang, B., & Huang, C. Z., (2015). Nano-toxicity of silver nanoparticles to red blood cells: Size dependent adsorption, uptake, and hemolytic activity. *Chemical Research in Toxicology, 28*(3), 501–509.
14. Cheng, Y., Su, H., Koop, T., Mikhailov, E., & Pöschl, U., (2015). Size dependence of phase transitions in aerosol nanoparticles. *Nature Communications, 6*, 9. Article ID: 5923.
15. Cheng, Y., Yin, L., Lin, S., Wiesner, M., Bernhardt, E., & Liu, J., (2011). Toxicity reduction of polymer-stabilized silver nanoparticles by sunlight. *The Journal of Physical Chemistry C, 115*(11), 4425–4432.
16. Christensen, F. M., Johnston, H. J., Stone, V., Aitken, R. J., Hankin, S., Peters, S., & Aschberger, K., (2010). Nano-silver-feasibility and challenges for human health risk assessment based on open literature. *Nanotoxicology, 4*(3), 284–295.

17. Cormode, D. P., Jarzyna, P. A., Mulder, W. J., & Fayad, Z. A., (2010). Modified natural nanoparticles as contrast agents for medical imaging. *Advanced Drug Delivery Reviews, 62*(3), 329–338.
18. Cushen, M., Kerry, J., Morris, M., Cruz-Romero, M., & Cummins, E., (2012). Nanotechnologies in the food industry-recent developments, risks and regulation. *Trends in Food Science & Technology, 24*(1), 30–46.
19. Dasarahalli-Huligowda, L. K., Goyal, M. R., & Suleria, H. A. R., (2019). In: Goyal, M. R., (ed.), *Nanotechnology Applications in Dairy Science: Packaging, Processing, and Preservation* (1st edn., p. 275). Oakville - ON: Apple Academic Press.
20. De Moura, M. R., Mattoso, L. H., & Zucolotto, V., (2012). Development of cellulose-based bactericidal nanocomposites containing silver nanoparticles and their use as active food packaging. *Journal of Food Engineering, 109*(3), 520–524.
21. Kumar, L. D. H., & Sarkar, P., (2018). Potential of nanotechnology in dairy processing: A review. Chapter 3. In: Goyal, M. R., (eds.), *Sustainable Biological Systems for Agriculture: Emerging Issues in Nanotechnology, Biofertilizers, Wastewater, Farm Machines* (pp. 55–80). Oakville - ON: Apple Academic Press.
22. Dickinson, E., (2012). Use of nanoparticles and microparticles in the formation and stabilization of food emulsions. *Trends in Food Science & Technology, 24*(1), 4–12.
23. Duncan, T. V., (2011). Applications of nanotechnology in food packaging and food safety: Barrier materials, antimicrobials and sensors. *Journal of Colloid and Interface Science, 363*(1), 1–24.
24. Elder, A., Yang, H., Gwiazda, R., Teng, X., Thurston, S., He, H., & Oberdörster, G., (2007). Testing nanomaterials of unknown toxicity: An example based on platinum nanoparticles of different shapes. *Advanced Materials, 19*(20), 3124–3129.
25. Espitia, P. J. P., Soares, N. D. F. F., Dos, R. C. J. S., De Andrade, N. J., Cruz, R. S., & Medeiros, E. A. A., (2012). Zinc oxide nanoparticles: Synthesis, antimicrobial activity and food packaging applications. *Food and Bioprocess Technology, 5*(5), 1447–1464.
26. Farhoodi, M., (2016). Nanocomposite materials for food packaging applications: Characterization and safety evaluation. *Food Engineering Reviews, 8*(1), 35–51.
27. Ghorbanzade, T., Jafari, S. M., Akhavan, S., & Hadavi, R., (2017). Nano-encapsulation of fish oil in nano-liposomes and its application in fortification of yogurt. *Food Chemistry, 216,* 146–152.
28. Green, R. J., Murphy, A. S., Schulz, B., Watkins, B. A., & Ferruzzi, M. G., (2007). Common tea formulations modulate *in vitro* digestive recovery of green tea catechins. *Molecular Nutrition & Food Research, 51*(9), 1152–1162.
29. Hajipour, M. J., Fromm, K. M., Ashkarran, A. A., De Aberasturi, D. J., De Larramendi, I. R., Rojo, T., Serpooshan, V., et al., (2012). Antibacterial properties of nanoparticles. *Trends in Biotechnology, 30*(10), 499–511.
30. Hao, L., Chen, L., Hao, J., & Zhong, N., (2013). Bioaccumulation and sub-acute toxicity of zinc oxide nanoparticles in juvenile carp (*Cyprinus carpio*): A comparative study with its bulk counterparts. *Ecotoxicology and Environmental Safety, 91,* 52–60.
31. Hassellöv, M., Readman, J. W., Ranville, J. F., & Tiede, K., (2008). Nanoparticle analysis and characterization methodologies in environmental risk assessment of engineered nanoparticles. *Ecotoxicology, 17*(5), 344–361.
32. Hati, S., Makwana, M. R., & Mandal, S., (2019). Encapsulation of probiotics for enhancing the survival in gastrointestinal tract. Chapter 9. In: Lohith, K. D. H.,

Goyal, M. R., & Suleria, H. A. R., (eds.), *Nanotechnology Applications in Dairy Science: Packaging, Processing, Preservation* (pp. 225–244). Oakville - ON: Apple Academic Press.

33. Holgate, S. T., (2010). Exposure, uptake, distribution and toxicity of nanomaterials in humans. *Journal of Biomedical Nanotechnology, 6*(1), 1–19.

34. Hong, T. K., Tripathy, N., Son, H. J., Ha, K. T., Jeong, H. S., & Hahn, Y. B., (2013). A comprehensive *in vitro* and *in vivo* study of ZnO nanoparticles toxicity. *Journal of Materials Chemistry B, 1*(23), 2985–2992.

35. Hristozov, D. R., Gottardo, S., Critto, A., & Marcomini, A., (2012). Risk assessment of engineered nanomaterials: A review of available data and approaches from a regulatory perspective. *Nanotoxicology, 6*(8), 880–898.

36. Hu, B., Pan, C., Sun, Y., Hou, Z., Ye, H., Hu, B., & Zeng, X., (2008). Optimization of fabrication parameters to produce chitosan-tripolyphosphate nanoparticles for delivery of tea catechins. *Journal of Agricultural and Food Chemistry, 56*(16), 7451–7458.

37. Jiang, J., Oberdörster, G., & Biswas, P., (2009). Characterization of size, surface charge, and agglomeration state of nanoparticle dispersions for toxicological studies. *Journal of Nanoparticle Research, 11*(1), 77–89.

38. Kandlikar, M., Ramachandran, G., Maynard, A., Murdock, B., & Toscano, W. A., (2007). Health risk assessment for nanoparticles: A case for using expert judgment. *Journal of Nanoparticle Research, 9*(1), 137–156.

39. Karakoti, A., Hench, L., & Seal, S., (2006). The potential toxicity of nanomaterials— the role of surfaces. *JOM, 58*(7), 77–82.

40. Kelly, K. L., Coronado, E., Zhao, L. L., & Schatz, G. C., (2003). The optical properties of metal nanoparticles: The influence of size, shape, and dielectric environment. *The Journal of Physical Chemistry B, 107*(3), 668–677.

41. Khalaj, M. J., Ahmadi, H., Lesankhosh, R., & Khalaj, G., (2016). Study of physical and mechanical properties of polypropylene nanocomposites for food packaging application: Nano-clay modified with iron nanoparticles. *Trends in Food Science & Technology, 51*, 41–48.

42. Kroll, A., Pillukat, M. H., Hahn, D., & Schnekenburger, J., (2009). Current *in vitro* methods in nanoparticle risk assessment: Limitations and challenges. *European Journal of Pharmaceutics and Biopharmaceutics, 72*(2), 370–377.

43. Kumar, D. L., Mitra, J., & Roopa, S., (2020). Nanoencapsulation of food carotenoids. Chapter 7. In: Ranjan, S. R., Dasgupta, N., & Eric, L., (eds.), *Environmental Nanotechnology* (Vol. 3, pp. 203–242). London: Springer.

44. Kumar, D. L., & Sarkar, P., (2017). Nanoemulsions for nutrient delivery in food. Chapter 4. In: Ranjan, S., Dasgupta, N., & Lichtfouse, E., (eds.), *Nanoscience in Food and Agriculture* (Vol. 5, pp. 81–121). Cham: Springer.

45. Le Corre, D., Bras, J., & Dufresne, A., (2010). Starch nanoparticles: A review. *Biomacromolecules, 11*(5), 1139–1153.

46. Lee, S. H., Heng, D., Ng, W. K., Chan, H. K., & Tan, R. B., (2011). Nano spray drying: A novel method for preparing protein nanoparticles for protein therapy. *International Journal of Pharmaceutics, 403*(1), 192–200.

47. Lohith, K. D. H., & Sarkar, P., (2018). Encapsulation of bioactive compounds using nanoemulsions. *Environmental Chemistry Letters, 16*(1), 59–70.

48. Lundqvist, M., Stigler, J., Elia, G., Lynch, I., Cedervall, T., & Dawson, K. A., (2008). Nanoparticle size and surface properties determine the protein corona with possible implications for biological impacts. *Proceedings of the National Academy of Sciences, 105*(38), 14265–14270.

49. McDougall, G. J., Dobson, P., Smith, P., Blake, A., & Stewart, D., (2005). Assessing potential bioavailability of raspberry anthocyanins using an *in vitro* digestion system. *Journal of Agricultural and Food Chemistry, 53*(15), 5896–5904.

50. Meyer, J., (1980). Gastric emptying of ordinary food: Effect of antrum on particle size. *American Journal of Physiology-Gastrointestinal and Liver Physiology, 239*(3), G133–G135.

51. Mohammed, F. A., Balaji, K., Girilal, M., Kalaichelvan, P., & Venkatesan, R., (2009). Mycobased synthesis of silver nanoparticles and their incorporation into sodium alginate films for vegetable and fruit preservation. *Journal of Agricultural and Food Chemistry, 57*(14), 6246–6252.

52. Mohanty, D., Jena, R., Choudhury, P. K., Pattnaik, R., Mohapatra, S., & Saini, M. R., (2016). Milk derived antimicrobial bioactive peptides: A review. *International Journal of Food Properties, 19*(4), 837–846.

53. Morabito, N., Crisafulli, A., Vergara, C., Gaudio, A., Lasco, A., Frisina, N., D'Anna, R., et al., (2002). Effects of genistein and hormone-replacement therapy on bone loss in early postmenopausal women: A randomized double-blind placebo-controlled study. *Journal of Bone and Mineral Research, 17*(10), 1904–1912.

54. Nie, Z., Petukhova, A., & Kumacheva, E., (2010). Properties and emerging applications of self-assembled structures made from inorganic nanoparticles. *Nature Nanotechnology, 5*(1), 15–25.

55. Niemeyer, C. M., (2003). Functional hybrid devices of proteins and inorganic nanoparticles. *Angewandte Chemie International Edition, 42*(47), 5796–5800.

56. Oberdörster, G., Oberdörster, E., & Oberdörster, J., (2005). Nanotoxicology: An emerging discipline evolving from studies of ultrafine particles. *Environmental Health Perspectives,* 823–839.

57. Pan, Y., Lin, Y., Liu, Y., & Liu, C., (2016). Size-dependent magnetic and electrocatalytic properties of nickel phosphide nanoparticles. *Applied Surface Science, 366,* 439–447.

58. Pelaz, B., Del, P. P., Maffre, P., Hartmann, R., Gallego, M., Rivera-Fernandez, S., De La Fuente, J. M., et al., (2015). surface functionalization of nanoparticles with polyethylene glycol: Effects on protein adsorption and cellular uptake. *ACS Nano, 9*(7), 6996–7008.

59. Peters, C. M., Green, R. J., Janle, E. M., & Ferruzzi, M. G., (2010). Formulation with ascorbic acid and sucrose modulates catechin bioavailability from green tea. *Food Research International, 43*(1), 95–102.

60. Podsędek, A., Redzynia, M., Klewicka, E., & Koziołkiewicz, M., (2014). Matrix effects on the stability and antioxidant activity of red cabbage anthocyanins under simulated gastrointestinal digestion. *BioMed Research International, 2014,* 11–18.

61. Raman, M., & Doble, M., (2019). Polyphenol nanoformulations for cancer therapy: Role of milk components. Chapter 6. In: Lohith, K. D. H., Goyal, M. R., & Suleria, H. A. R., (eds.), *Nanotechnology Applications in Dairy Science: Packaging, Processing, Preservation* (pp. 123–168). Oakville - ON: Apple Academic Press.

62. Rao, C., Kulkarni, G., Thomas, P. J., & Edwards, P. P., (2002). Size-dependent chemistry: Properties of nanocrystals. *Chemistry-A European Journal, 8*(1), 28–35.
63. Rao, J. P., & Geckeler, K. E., (2011). Polymer nanoparticles: Preparation techniques and size-control parameters. *Progress in Polymer Science, 36*(7), 887–913.
64. Rico, C. M., Majumdar, S., Duarte-Gardea, M., Peralta-Videa, J. R., & Gardea-Torresdey, J. L., (2011). Interaction of nanoparticles with edible plants and their possible implications in the food chain. *Journal of Agricultural and Food Chemistry, 59*(8), 3485–3498.
65. Robichaud, C. O., Uyar, A. E., Darby, M. R., Zucker, L. G., & Wiesner, M. R., (2009). Estimates of upper bounds and trends in nano-TiO_2 production as a basis for exposure assessment. *Environmental Science & Technology, 43*(12), 4227–4233.
66. Sarkar, P., Lohith, K. D. H., Dhumal, C., Panigrahi, S. S., & Choudhary, R., (2015). Traditional and ayurvedic foods of indian origin. *Journal of Ethnic Foods, 2*(3), 97–109.
67. Satoh, N., Nakashima, T., Kamikura, K., & Yamamoto, K., (2008). Quantum size effect in TiO_2 nanoparticles prepared by finely controlled metal assembly on dendrimer templates. *Nature Nanotechnology, 3*(2), 106–111.
68. Savolainen, K., Alenius, H., Norppa, H., Pylkkänen, L., Tuomi, T., & Kasper, G., (2010). Risk Assessment of engineered nanomaterials and nanotechnologies: A review. *Toxicology, 269*(2), 92–104.
69. Sharma, V. K., (2009). Aggregation and toxicity of titanium dioxide nanoparticles in aquatic environment: A review. *Journal of Environmental Science and Health Part A, 44*(14), 1485–1495.
70. Sharma, V. K., Yngard, R. A., & Lin, Y., (2009). Silver nanoparticles: Green synthesis and their antimicrobial activities. *Advances in Colloid and Interface Science, 145*(1), 83–96.
71. Shipway, A. N., & Willner, I., (2001). Nanoparticles as structural and functional units in surface-confined architectures. *Chemical Communications*, (20), 2035–2045.
72. Silvestre, C., Duraccio, D., & Cimmino, S., (2011). Food packaging based on polymer nanomaterials. *Progress in Polymer Science, 36*(12), 1766–1782.
73. Song, J. Y., & Kim, B. S., (2009). Rapid biological synthesis of silver nanoparticles using plant leaf extracts. *Bioprocess and Biosystems Engineering, 32*(1), 79–84.
74. Sowbhagya, H. B., Lohith, K. D. H., Anush, S. M., & Jagan, M. R. L., (2015). Microwave impact on the flavor compounds of cinnamon bark (*Cinnamomum Cassia*) Volatile oil and polyphenol extraction. *Current Microwave Chemistry, 2*, 51–58.
75. Sozer, N., & Kokini, J. L., (2009). Nanotechnology and its applications in the food sector. *trends in biotechnology, 27*(2), 82–89.
76. Subbiah, R., Veerapandian, M., & Yun, K. S., (2010). Nanoparticles: Functionalization and multifunctional applications in biomedical sciences. *Current Medicinal Chemistry, 17*(36), 4559–4577.
77. Tiede, K., Boxall, A. B., Tear, S. P., Lewis, J., David, H., & Hassellöv, M., (2008). Detection and characterization of engineered nanoparticles in food and the environment. *Food Additives and Contaminants, 25*(7), 795–821.
78. Tsuji, J. S., Maynard, A. D., Howard, P. C., James, J. T., Lam, C. W., Warheit, D. B., & Santamaria, A. B., (2006). Research strategies for safety evaluation of nanomaterials, part IV: Risk assessment of nanoparticles. *Toxicological Sciences, 89*(1), 42–50.

79. Voinov, M. A., Pagán, J. O. S., Morrison, E., Smirnova, T. I., & Smirnov, A. I., (2010). Surface-mediated production of hydroxyl radicals as a mechanism of iron oxide nanoparticle biotoxicity. *Journal of the American Chemical Society, 133*(1), 35–41.

80. Wang, H., Wick, R. L., & Xing, B., (2009). Toxicity of nanoparticulate and bulk ZnO, Al_2O_3 and TiO_2 to the nematode *Caenorhabditis elegans*. *Environmental Pollution, 157*(4), 1171–1177.

81. Wang, Y., & Xia, Y., (2004). Bottom-up and top-down approaches to the synthesis of monodispersed spherical colloids of low melting-point metals. *Nano Letters, 4*(10), 2047–2050.

82. Weir, A., Westerhoff, P., Fabricius, L., Hristovski, K., & Von, G. N., (2012). titanium dioxide nanoparticles in food and personal care products. *Environmental Science & Technology, 46*(4), 2242–2250.

83. Wijnhoven, S. W., Peijnenburg, W. J., Herberts, C. A., Hagens, W. I., Oomen, A. G., Heugens, E. H., Roszek, B., et al., (2009). Nano-silver: A review of available data and knowledge gaps in human and environmental risk assessment. *Nanotoxicology, 3*(2), 109–138.

84. Wong, S. W., Leung, P. T., Djurišić, A., & Leung, K. M., (2010). Toxicities of nano zinc oxide to five marine organisms: Influences of aggregate size and ion solubility. *Analytical and Bioanalytical Chemistry, 396*(2), 609–618.

85. Wu, F., Liu, D., Wang, T., Li, W., & Zhou, X., (2015). Different surface properties of L-Arginine functionalized silver nanoparticles and their influence on the conductive and adhesive properties of nanosilver films. *Journal of Materials Science: Materials in Electronics, 26*(9), 6781–6786.

86. Yah, C. S., Simate, G. S., & Iyuke, S. E., (2012). Nanoparticles toxicity and their routes of exposures. *Pakistan Journal of Pharmaceutical Sciences, 25*(2), 477–491.

87. Zänker, H., & Schierz, A., (2012). Engineered nanoparticles and their identification among natural nanoparticles. *Annual Review of Analytical Chemistry, 5*, 107–132.

CHAPTER 6

NANOTECHNOLOGY IN PROCESSING AND PRESERVATION OF MEAT AND MEAT PRODUCTS

SOUMITRA BANERJEE, H. B. MURALIDHARA, and PREETAM SARKAR

ABSTRACT

Nowadays with advancement in production, processing, packaging, there is an increase in the demand of processed meat products in terms of taste and sensory qualities, extended shelf-life, digestibility, nutrition retention, antimicrobial properties, fortification of essential nutrients, etc. The improved and smarter packaging has also been a boon to the meat industry at the global level. In meat processing sectors, Nanotechnology has wide applicability from processing, packaging, improving the nutritional quality to extending shelf-life. Though being a new technology, yet many times hurdles are often encountered in terms of food safety and security factors. The aim of this chapter is to review the potential applications of nanotechnology in meat processing and preservation, its recent developments, and limitations.

6.1 INTRODUCTION

Animal-based foods serve as a good source of nutrition, such as: full of energy, protein, and micronutrients (e.g.: iron, zinc, vitamin A, vitamin B-12), etc. [12, 23]. With time, primitive humans had understood the concept of easy digesting properties of roasted or cooked, processed, and preserved meat. Similar to primitive human history to modern human history, meat, and meat products have occupied important role in our daily diet. Meat available in the processed form is a good source of nutrition and is also preferred as a delicacy. Almost all countries of the world consume meat and meat-based products. Farm domestic animals, on the other hand, can convert the roughages into human foods, like milk and meat. Pricing of the different types of meat is based on this conversion scale [32]. Following example for conversion scale from grain to meat is given by Potter et al. [32]:

Type of Meat	Amount of Grain Consumed/kg of Meat
Chicken	2
Pork	4
Beef	8

According to FAO World Food Outlook [16], meat production in the year 2012, 2013, 2014, and 2020 was 304.2, 308.5, 311.8 and 334 million metric tons, respectively. Change in pattern of food habits has been a key contributor to the rise in global meat consumption. With the rise in income and nutritional awareness, there is a rise in demand for livestock products like meat. In today's time, there is a growing trend in meat production and consumption. FAO [17] predicted about a double digit increase in the meat production by the year 2050, with the major contribution from the developing countries. Meat consumption pattern was well explained and predicted by Heinz and Hautzinger [19], where they reported that that the average annual per head per capita meat consumption was [17, 19]:

• 10 kg in the year 1960s;
• 26 kg in the year 2000s; and
• 37 kg by the year 2030.

According to FAO [17], meat is considered as a nutritional food source, containing high-quality protein and all essential amino acids, lipids, and fatty acids, essential micronutrients (such as: vitamins, minerals, other bioactive compounds) but with least amount of carbohydrates. Bioavailability of mineral and vitamins are high in meat and meat products.

Meat processing happened from historic times though recent developments particularly in European countries, where meat is successfully processed, for example: burger patties, frankfurter-type sausages, and cooked ham. Asian countries are also not lagging behind in the development of meat processing technologies, but there is less demand for processed meat in Asian and African countries [19]. With globalization and changes in lifestyle, consumers are slowly picking up with consumption of processed meat, particularly the younger generations.

Meat processing consists of different steps and procedures by treating them with different physical or chemical treatments. Modern technological developments have helped meat processing sectors to leap steps forward. One such particular application field is nanotechnology, which has many applications in meat processing and other fields [6, 8]. According to Bhushan [6], nanotechnology is the application, production, modification of any material at nano-scale ($<1 \times 10^{-9}$ m) [11]. This field is interdisciplinary in nature and finds wide applications from science to engineering. Food and bioprocessing sectors also find wide applications of nanotechnology for processing and producing foods with higher quality, safer in nature and with extended shelf-life, through top-down and bottom-up methods [6, 29, 34, 39].

Areas of applications of nanotechnology in food processing sectors are listed below [9, 35]:

- Antimicrobial packaging;
- Laminates and packaging materials;
- Nanoadditives;
- Nanocoatings or edible films;
- Nanodispersions and nanocapsules;
- Nanoemulsions for fat reduction;
- Nanoparticles;
- Self-sanitizing surfaces.

Nanotechnology increases the product quality and choices of food products, but there are issues about the safety concerns, e.g., deliberate, or accidental usage of nanoparticles (NPs) in food more than the recommended level. Besides this, Amini et al. [2] reviewed the harmful effects on human health and environmental conditions because of usage of nanomaterials in food processing and allied sectors.

Health risk assessment should be made mandatory for developed food products using nanotechnology. Also, it was recommended that there should be a better understanding of the size, dosage, surface chemistry and structures of nano-particles including nano-toxicology aspect, to develop safe food products [2].

This chapter discusses in details some possible technological applications of nanotechnology in meat and meat product processing with their advantages over the conventional methods of processing.

6.2   MEAT PROCESSING: BRIEF INTRODUCTION

Commonly meat comes from the flesh of terrestrial animals, such as: cattle (beef), calves (veal), hogs, sheep, dogs, kangaroos, etc. Meat products not only consist of only meat but also various byproducts, such as, intestine, animal fat, blood, bones, hormones, enzymes, etc., for use in food industries, cattle feed preparation and other pharmaceutical sectors. Meat processing consists of physical and chemical treatments, which commonly include following operations [17, 32, 44]:

1. **Size Reduction:** It consists of cutting and chopping of the larger sized meat chunk into smaller pieces. Types of equipment used for this operation are: the grinder, bowl cutter/chopper, knives, and choppers, etc.

2. **Ageing and Tenderizing:** After slaughtering, if the carcass of the dead animal is kept as such, then the flesh becomes tough due to rigor mortis. For preventing this, the slaughtered animal carcass may be kept under refrigerated conditions, and the muscles become tender. Tenderization may also be done by enzymatic action (i.e., papain). Aging or ripening is done by storage for a period of 1–4 weeks under controlled humidity conditions. UV light may be applied to prevent microbial growth at higher temperature tenderization. Artificially tenderization may be done by mechanical, electrical stimulations, etc.

3. **Curing:** It is different from aging or ripening operation, where the intention is to enhance the shelf-life, flavor, color, and tenderness by adding curing ingredients. Curing is quite an old process of meat preservation. Commonly used ingredients for meat curing are sodium chloride, sodium nitrate, sugar, spices, etc. Curing agents are mixed in proper proportion and are available commercially as curing mixtures or may be formulated by the meat processors by mixing the ingredients.

4. **Smoking:** Cured meats are smoked, which is also a method of mild preservation to develop a typical desirable smoky flavor. Smoke

is generated by burning hardwood logs or chips, over which the meat is kept in hanging position for a period of 18–24 hours at 50–52°C. This is the traditional method of smoking meat; whereas, in modern days, processors use smoke room or tunnel. Required smoke can be generated even by developed electrical methods or by liquid smoke, which is basically the flavor chemicals extracted from the smoke.

5. **Packaging and Storage:** Freezing and storage of meat is a good option for meat preservation, where meat can be kept even for years. Consumers prefer to have de-frozen meat, which is kept under the refrigerated conditions at home, but restaurants and military have more demand for frozen hassle-free meat. Vacuum packaging is a good option for storing beef and pork, for their safe storage at 0°C for about 3 weeks. Modified atmospheric packaging (MAP) eliminates oxygen inside the package, thus decreases chances of meat spoilage. High barrier films are being utilized to reduce/prevent moisture and oxygen migration from the atmosphere to the sealed packaging environment.

6. **Cooking:** It is done to make the meat digestible after consumption. During cooking, the meat becomes tender because of melting of fats, dissolution of connective collagen, separation of muscle fibers, etc. Cooking of meat involves thermal treatment, for getting desirable texture, flavor, color, and also making meat safe from microbial contamination. Similar to other food products, thermal treatments of meat may be classified as pasteurization and sterilization. Cooking of meat may be done at low temperature for longer time, which makes the meat tender and soft. Higher temperature cooking like frying or grilling produces heterocyclic amines (HCA) and polycyclic aromatic hydrocarbons (PAH), which may indicate carcinogenic effects.

Modern-day meat processors follow the conventional processing, besides which they work on the development of new processing methods for meat processing and new products. Modern day's consumers not only focus on the taste of the product but also on its nutrition and health effects. Nanotechnology plays a significant role in the development of processing, preservation, and quality of the meat products. This technology has proven its availability to provide healthier and quality food products at economical prices. Meat processing also has a wide array of applications in nanotechnology.

6.3 APPLICATIONS OF NANOTECHNOLOGY IN MEAT PROCESSING AND PRESERVATION (Figure 6.1)

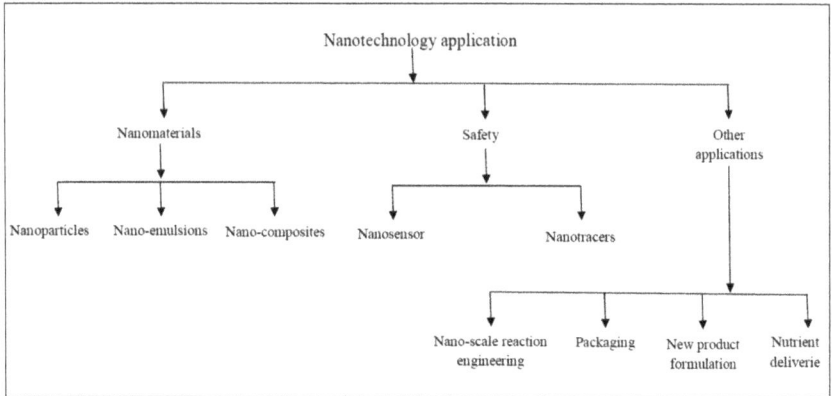

FIGURE 6.1   Nanotechnology applications in meat processing.

6.3.1 NANOMATERIALS

Nanomaterials or NPs are used in meat processing for many purposes, such as: for better color/flavor/odor and flow properties of meat/ meat products after processing and for an extended shelf-life. Plastic films for packaging meat materials are made from NPs (such as: silicate NPs, ZnO, and titanium oxide) to prevent moisture permeation. Silver nanoparticles (AgNPs) prevent microbial growth due its antimicrobial property [33]. Nano-sized nutritional supplements and nutraceuticals particles present in developed meat products have the capacity to enhance the taste, absorption, and better bioavailability. NPs are capable of active ingredients delivery and stabilizing bioactive compounds [5, 22].

In the meat industries, there is a requirement for development of modified additives in the form of NPs to replace the existing additives due to potential health hazards. Nutritional profile of meat is quite high due to high amount of fats and proteins along with other essential micronutrients (such as: minerals), though the bioavailability of the bioactive compounds is quite low. Nanotechnology may be used wisely as a process-based innovation to address these problems and assure the consumers on the food safety. There is a rising demand for meat products with the low level of sodium salts, fats, cholesterol, nitrites, and calorific values, but with high amounts of bioactive compounds and fiber, etc., without compromising the taste and other sensory attributes of meat.

Various approaches for potential and functional meat products development have been reported, such as: modification of lipid profile, salt, and nitrites reduction, and utilization of naturally occurring antimicrobials in nanoemulsion forms [22, 30, 47]. Nano-powder forms were found to be more effective than the conventional form, i.e., ginger micro-sized powder when applied was found to have much greater penetrability and tenderizing effects in comparison to raw form, with much high surface area causing increase in fluidity, water holding capacity, water solubility index and protein solubility.

Potential of nanopowder is well-recognized, but there is a need to understand the change in behavioral properties of the powder, nature of raw materials, characteristics, etc. [51], which would have potential applications in meat processing sectors. A similar study was conducted on wheat bran dietary fiber, which was subjected to ultrafine grinding. The ultrafine ground dietary fiber was found to have reduced hydration properties but was found to have better functional properties and higher dispersibility and solubility in the food system [52]. Properties of nanomaterials are often found to vary widely from their native form [22].

6.3.2 NANOEMULSIONS

Emulsions have its wide applications in food sectors that are often found to be unstable in its own form. Nanoemulsion may be formed by high-pressure homogenization or by catastrophic inversion. After homogenization, smaller droplets often have the tendency to clump together and form large droplets, which may be prevented by encapsulation process to prevent the recombination. Nanoemulsion stabilization is also achievable by electrostatic or steric stabilization or by increasing the emulsion viscosity. In the food sector, nanoemulsion finds its application in functional foods and beverages, novel encapsulation, and smart sensing systems [39].

Salminen et al. [37] studied oil-in-water emulsion of n-3 fatty acids incorporation in meat products. Physically the emulsion was found stable but its incorporation in meat matrix caused reduced oxidative stability [37]. Antimicrobial activity of essential oils (such as: oregano, thyme, basil, marjoram, lemongrass, ginger, and clove oil) was reported for minced meat preparation, where particularly foodborne pathogens and spoilage-causing bacteria were investigated. The effectiveness varied with experimentally and naturally present microorganisms in the meat sample [4]. These kind of natural antimicrobial agents for meat preservation are nowadays getting more interest and nanoemulsion form is gaining popularity as a method of controlled delivery of active ingredients in meat products.

The microbial activity of microfluidized lemon grass oil-alginate nanoemulsion was tested against *E. coli*, and its bactericidal effects were observed [38]. Another study was reported on the application of nanoemulsion prepared with biocompounds extracted from sunflower oil; and it was found to have efficient antimicrobial effects on meat-borne pathogens. This work shows possibilities of replacing chemical preservatives with safer emulsions [7]. Nanoemulsions were not only found effective against a variety of meat pathogens but were found effective for surface sterilization of packaging materials and other equipment [42] in the meat industry.

6.3.3   NANOENCAPSULATION

The nanoencapsulation method [22, 24] increases the bioavailability of bioactive compounds and helps in their controlled release at the specific target site. Based on the requirement, nanoemulsions may be applied directly as liquid state or may be applied in spray-dried powder form. NPs may be produced by high shear stirring, high-speed or high-pressure homogenizers, ultrasonicator, and microfluidizer [15, 28].

Essential oils have antimicrobial properties that play significant role to reduce spoilage and pathogenic microorganisms. This property can be used as the alternative to chemical preservatives for extending the shelf-life of meat and meat products. Moraes-Lovison et al. [26] reported their work on the production of nanoemulsion encapsulation of oregano essential oil using phase inversion temperature method; and the emulsion showed good antibacterial activity. The process technology was found suitable as an alternate of chemical preservatives for the preservation of chicken pate and shelf-life extension. In another study, encapsulated thyme essential oil in chitosan NPs showed better antimicrobial and antioxidant activities under the refrigerated conditions in beef burgers. Addition of encapsulated thyme essential oil resulted in extended shelf-life and least vaporization of bioactive compounds during storage [18]. A similar kind of study reported on the use of nano-encapsulated forms of *Zataria Multiflora* essential oil on beef burger, which extended the shelf-life of the burger under refrigeration conditions [45].

6.3.4   NANOSENSORS

Nanosensors are used for rapid detection of pathogens and spoilage causing microbes or other undesirable agents. For monitoring the packaged meat quality, these sensors are added to the packaging material to monitor the food quality. For example, CO_2 enriched MAP for fresh raw bacon changed the color with the increase in oxygen content inside the package space [25].

Unique NPs may be devised efficient enough to detect the foul odor, chemical contamination, pathogens, or inside packaging atmospheric changes. These changes would alert the consumers. These nanosensors would be helpful for the consumers [11]. Nanosensors can readily detect spoilage in meat at the initial stages, which is helpful for the maintaining the quality of the meat products. Early spoilage detection is helpful to prevent further spoilage by taking recommended steps. It is also possible to detect the presence of pathogens in meat nanosensors. Though the field is exciting and has wide application domain, yet challenges are faced for commercial-scale applications [14, 22].

Development of nanosensors for ensuring food safety, smart food packaging and biosecurity are given as important application of nanotechnology [1, 27, 31, 36]. Carbon nanotubes (CNTs) are tube-shaped structures made of carbon, with nanoscale diameter. Optical immunosensor based on single-walled CNT was utilized for efficient detection of Staphylococcal enterotoxins [50]. Different authors have reviewed different methods of detection of organic molecules, gases, microorganisms using nanosensors [3, 11, 43].

Nanosensors find its applicability in storage rooms to detect changes in temperature, relative humidity, microbial contamination, storage product degradation, gas levels, etc., and respond accordingly. For meat products, nanosensors may be used for determining the "Freshness Index" and also the oxygen content of the storage condition.

Nanosensors find its application in "Electronic Tongue" technology, which uses a series of sensors to detect gases release by the contaminating organisms or other unwanted changes in meat, etc. Nanotechnology based biosensors are capable to detect allergen protein present in foods and help the consumers to decide safer foods for consumption. Nanosensors would find its future applications in intelligent solutions, such as: tracking, tracing, and detecting the real-time quality of food products for food safety applications [34, 41]. Nano-tracers are helpful in determination of risks and hazards in meat processing plants [22].

6.3.5 PACKAGING

The nanotechnology is playing significant role in packaging film manu-facturing. The basic purpose of food packaging is to contain and protect the meat product, but the application of nanotechnology has changed the definition of packaging.

Now nano-packaging is being used for extending the shelf-life, reduction of gases and moisture permeability, and can act as antioxidant and antimicrobial protective covering over the product. Nanopackaging materials use nanocomposites, made of fillers like silicates, cellulose microfibrils, CNTs, etc. These types of materials have good barrier prop-erties, such as: oxygen barrier, UV blocking ability and better rigidity.

With the help of nanotechnology, it is possible to develop packaging materials with the capability to discharge antimicrobial, antioxidants, nutraceuticals to the packaged product to increase the shelf-life and the value of the food. It was possible to eliminate the trend of applying a waxy coating over the foods, and instead nano-scale edible coatings were developed. This nano-scale coating would find its applications not only for fruits but also for meats. Packaging material with inbuilt nano-sensors would be helpful in monitoring inside and outside conditions of the pack-aging materials. Nanotechnology-enabled radio frequency identification (RFID) tags are portable, smaller in size and even possible to be printed over packaging material or on the thin level, which is found helpful to track the product, respond to microbial contamination, extends shelf-life, helpful for anti-counterfeiting and provides temperature measurement.

Nanobiodegradable packaging is another eco-friendly future promising alternative to conventional plastics and other non-degradable laminate packaging materials. Polyamide plastic film with clay NPs packaging materials used to pack meat have resulted in longer shelf-life than conven-tionally packed meat, because of better oxygen, moisture, and CO_2 barrier properties [1, 22].

Yang et al. [49] studied nano-packaging material that was made from polyethylene with nano-powders of nano-Ag, kaolin, anatase TiO_2, and rutile TiO_2. The developed packaging material was able to increase the shelf-life of strawberries stored at 4°C, causing minor changes in sensory, composition, and physiological qualities as compared to the strawberries packaged in normal polyethylene pouch package under same storage conditions. Nanolaminates are helpful for vast range of foods from fruits and vegetables to meat, where nanolaminates may serve a barrier to moisture, lipids, gasses, etc. They may also be loaded with functional compounds, such as: preservatives, antioxidants, nutrients, antimicrobials, etc. [48].

6.4   REGULATION GUIDELINES FOR NANOTECHNOLOGY APPLICATIONS IN MEAT PROCESSING

It is essential to have a set of food safety and regulation guidelines, which needs to be followed for production of safe and quality foods that must meet standards of national and international markets. With development of better food inspection protocols and on-time surveillance systems, we have better control of food quality and prevention of food-related hazards. Food hazards may be recognized as: microbial, pesticide residue, harmful chemical contaminations, undesirable adulteration, etc. Introduction of hazard analysis critical control point (HACCP) has helped industries and processing units to prevent food hazards. To prevent contamination and assure quality production, following programs are used at various stages from production to procurement:

- Good agricultural practices (GAP);
- Good manufacturing practices (GMP); and
- Sanitation standard operating procedures (SSOP).

Codex Alimentarius Commission (CAC) serves as an international monitoring body and it regulates food quality and safety of foods based on the Food and Agriculture Organization (FAO) of the United Nations and the World Health Organization (WHO). The CAC is responsible for developing international standards for range of food products like pesticide residues, additives, labeling, etc. CAC is also involved in risk assessment for prevention of microbial hazards in foods.

Food regulatory authorities of different nations and different regions are different. Its purpose is to protect the consumer's health, fair food trade and promotion of all food standards for all agencies, including international, governmental, and non-governmental organizations.

In India, National codex contact point (NCCP) is *Codex India*, which involves in the promotion of *Codex* activities in India. Food safety management system (FSMS) under International Organization for Standardization (ISO 22000) addresses food safety management to monitor and control food hazards. In India under Government of India (GOI), there is Food Safety and Standards Act of 2006, which is responsible for implementation of food Laws and establishment of standards, regulations, and ensuring the availability of safe food products to the consumers [20, 21, 40].

6.4.1   REGULATIONS FOR MEAT AND FOOD PROCESSING

America's Federal Meat Inspection Act of 1906 (FMIA) helps to monitor adulteration and misbranding of meat and meat products. Poultry Products Inspection Act (PPIA, 1957) under United States Department of Agriculture (USDA) monitors the poultry products. Other livestock and poultry items not under FMIA and PPIA are covered by Food, Drug, and Cosmetic Act under the Food and Drug Administration (FDA). FDA is an active Federal agency of United States Department of Health and Human Services (US HHS), which protects and promotes public health by monitoring, supervision, and controlling foods and food products.

Food Safety and Inspection Service (FSIS-an agency of the USDA) monitors and regulates commercial supply of meat, poultry, and egg products for safety, labeling, and packaging. FSIS ensures authority from FMIA, PPIA, and Egg Products Inspection Act (EPIA, 1970); and is responsible for meat, poultry, and egg products imported to the United States. FSIS plays important role in prevention of food hazards and illness and works closely with federal, state, and local food safety communities.

In India, Meat Food Products Order (MFPO, 1973) under the Essential Commodities Act of 1955 by GOI included standards and limits for meat products. MFPO's important role is to protect the consumers of meat products by ensuring the food safety. Under the initiative of Government of India (GOI), Indian Standard Institution (ISI) was established for promotion and adaptation of standards throughout India. ISI formed Meat and Meat Products Sectional Committee, under Agricultural and Food Products Division Council (1958) for standardization of meat industry in India. These standards assist in the improvement of food safety, losses, and prevention of microbial hazards. FSSAI (2006) is also involved in certifying, monitoring, and regulation of food safety of foods and food products including meat [20, 32, 40].

6.4.2 NANOTECHNOLOGY AND FOOD REGULATIONS

By understanding how organics interact at nano level, the food scientists now have better understanding of nanotechnology and precise control over the process. One major drawback in commercial adaptability of nanotechnology is the safety issues and need for development of standards for process and product monitoring.

There is immediate requirement for governmental regulatory bodies for providing regulation and assessment of risks associated with nanotechnology applications in food sectors. European Union (EU) and United States Food and Drug Administration (US FDA) has their own sets of laws and regulations for nanotechnology applications in foods. US FDA regulates and monitors number of consumable products, such as, foods, cosmetics, veterinary products, etc., which use nanotechnology or NPs [46].

EU regulates foods, cosmetics, etc., for safer products for the use of nanomaterials. Joint Research Center (JRC) under EU furnishes scientific and technical guidance on execution and monitoring nanomaterials that are present in consumers products, such as: foods and cosmetics [13].

In Australia, nanoadditives in foods and other nanotechnology applications come under Food Standards Australia and New Zealand (FSANZ). In addition to the advances in nanotechnology application in foods, its regulatory considerations (such as: safety and risk assessments, toxicity, long-term effects on consumers health, environmental impact, economical aspect, and consumers acceptance) must be assessed and considered [1, 11, 31].

6.5 SUMMARY

Nanotechnology has huge potential in processing/preservation of meat and meat products. Research studies have shown that NPs, nanoemulsions, and nanoencapsulation technologies have wide applications in meat processing and preservation. Nanotechnology also found its application in the development of packaging materials and nano-sensors for preventing early spoilage of meat products. Before making the food products available in the market, there is a requirement to ensure strict vigilance over nano-technology applications for commercial scale food processing, to avoid any potential health hazards. The consumers also need to be educated about the food product benefits and their safety aspects.

ACKNOWLEDGMENT

Authors are grateful to "Center for Incubation, Innovation, Research, and Consultancy (CIIRC), Jyothy Institute of Technology-Bengaluru (Karnataka, India)" and "National Institute of Technology-Rourkela (Orissa, India)" for providing all necessary facilities and infrastructure for successful compilation of this chapter.

KEYWORDS

- **food safety**
- **meat preservation**
- **meat processing**
- **nanoemulsion**
- **nanosensors**
- **nanotracers**
- **regulations**

REFERENCES

1. Lfadul, S. M., & Elneshwy, A. A., (2010). Use of nanotechnology in food processing, packaging and safety-review. *African Journal of Food, Agriculture, Nutrition and Development, 10*(6), 2719–2739.
2. Amini, S. M., Gilaki, M., & Karchani, M., (2014). Safety of nanotechnology in food industries. *Electronic Physician, 6*(4), 962–970.
3. Baltić, M. Ž., Bošković, M., Ivanović, J., Dokmanović, M., Janjić, J., Lončina, J., & Baltić, T., (2013). Nanotechnology and its potential applications in meat industry. *Tehnologija Mesa, 54*(2), 168–175.

4. Barbosa, L. N., Rall, V. L. M., Fernandes, A. A. H., Ushimaru, P. I., Da Silva, P. I., & Fernandes, Jr. A., (2009). Essential oils against foodborne pathogens and spoilage bacteria in minced meat. *Foodborne Pathogens and Disease, 6*(6), 725–728.

5. Berekaa, M. M., (2015). Nanotechnology in food industry: Advances in food processing, packaging and food safety. *International Journal of Current Microbiology and Applied Science, 4*(5), 345–357.

6. Bhushan, B., (2010). Introduction to nanotechnology. *Springer Handbook of Nanotechnology* (Vol. 1, pp. 1–6) Springer Science & Business Media.

7. Bradeeba, K., & Sivakumaar, P. K., (2013). Effects of the component of sunflower oil based nanoemulsions against meat-borne pathogens. *International Journal of Applied Biology and Pharmaceutical Technology, 4*(1), 88–95.

8. Charles, P., & Frank, O., (2003). *Introduction to Nanotechnology* (pp. 1–7). New York: John Wiley & Sons, Inc.

9. Chellaram, C., Murugaboopathi, G., John, A. A., Sivakumar, R., Ganesan, S., Krithika, S., & Priya, G., (2014). Significance of nanotechnology in food industry. *APCBEE Procedia, 8*, 109–113.

10. Delgado, C. L., (2003). Rising consumption of meat and milk in developing countries has created a new food revolution. *The Journal of Nutrition, 133*(11), 3907S–3910S.

11. Duncan, T. V., (2011). Applications of nanotechnology in food packaging and food safety: Barrier materials, antimicrobials and sensors. *Journal of Colloid and Interface Science, 363*(1), 1–24.

12. Eaton, S. B., (2006). The ancestral human diet: What was it and should it be a paradigm for contemporary nutrition. *Proceedings of the Nutrition Society, 65*(1), 1–6. doi: 10.1079/PNS2005471.

13. European Union (EU), (2017). *Nanotechnology.* https://ec.europa.eu/jrc/en/research-topic/nanotechnology (accessed on 06 October 2021).

14. Evans, H. M., (2009). Nanotechnology-enabled sensing. *Report of the National Nanotechnology Initiative Workshop*, 27–34.

15. Ezhilarasi, P. N., Karthik, P., Chhanwal, N., & Anandharamakrishnan, C., (2013). Nanoencapsulation techniques for food bioactive components: A review. *Food and Bioprocess Technology, 6*(3), 628–647.

16. FAO World Food Outlook, (2014). *Meat Consumption.* Rome: Food and Agriculture Organization of the United Nations. http://www.fao.org/ag/againfo/themes/en/meat/background.html (accessed on 05 October 2021).

17. FAO (Food and Agriculture Organization of the United Nations), (2016). *Meat and Meat Products.* http://www.fao.org/ag/againfo/themes/en/meat/home.html (accessed on 05 October 2021).

18. Ghaderi-Ghahfarokhi, M., Barzegar, M., Sahari, M. A., & Azizi, M. H., (2016). Nanoencapsulation approach to improve antimicrobial and antioxidant activity of thyme essential oil in beef burgers during refrigerated storage. *Food and Bioprocess Technology, 9*(7), 1187–1201.

19. Heinz, G., & Hautzinger, P., (2007). *Meat Processing Technology for Small to Medium Scale Producers* (p. 56). Rome: Food and Agriculture Organization of the United Nations, RAP (Regional Office for Asia and the Pacific) Publication # 2007/20.

20. Indian Council of Agricultural Research (ICAR), (2012). *Handbook of Agricultural Engineering, Directorate of Knowledge Management in Agriculture* (p. 738). India.

21. Jukes, D., (2009). Quality assurance and legislation. Chapter 15. In: Campbell-Platt, G., (ed.), *Food Science and Technology* (p. 353). London: Wiley Blackwell-A John Wiley & Sons Ltd.

22. Khan, M. I., Sahar, A., & Rahman, U., (2016). Nanotechnology in healthier meat processing. In: Grumezescu, A. M., (ed.), *Novel Approaches of Nanotechnology in Food* (Vol. 1, pp. 313–357). London: Academic Press, Elsevier.

23. Klein, R. G., (1999). *The Human Career: Human Biological and Cultural Origins* (p. 183). Chicago, IL: The University of Chicago Press.

24. Konan, Y. N., Gurny, R., & Allémann, E., (2002). Preparation and characterization of sterile and freeze-dried sub-200nm nanoparticles. *International Journal of Pharmaceutics, 233*(1), 239–252.

25. Mills, A., (2005). Oxygen indicators and intelligent inks for packaging food. *Chemical Society Reviews, 34*(12), 1003–1011.

26. Moraes-Lovison, M., Marostegan, L. F., Peres, M. S., Menezes, I. F., Ghiraldi, M., Rodrigues, R. A., Fernandes, A. M., & Pinho, S. C., (2017). Nanoemulsions encapsulating oregano essential oil: Production, stability, antibacterial activity and incorporation in chicken pâté. *LWT-Food Science and Technology, 77*, 233–240.

27. Moraru, C. I., Panchapakesan, C. P., Huang, Q., Takhistov, P., Liu, S., & Kokini, J. L., (2003). Nanotechnology: A new frontier in food science. *Food Technology, 57*(12), 24–29.

28. Mozafari, M. R., Flanagan, J., Matia-Merino, L., Awati, A., Omri, A., Suntres, Z. E., & Singh, H., (2006). Recent trends in the lipid-based nanoencapsulation of antioxidants and their role in foods. *Journal of the Science of Food and Agriculture, 86*(13), 2038–2045.

29. Neethirajan, S., & Jayas, D. S., (2011). Nanotechnology for the food and bioprocessing industries. *Food and Bioprocess Technology, 4*(1), 39–47.

30. Olmedilla-Alonso, B., Jiménez-Colmenero, F., & Sánchez-Muniz, F. J., (2013). Development and assessment of healthy properties of meat and meat products designed as functional foods. *Meat Science, 95*(4), 919–930.

31. Ozimek, L., Pospiech, E., & Narine, S., (2010). Nanotechnologies in food and meat processing. *ACTA Scientiarum Polonorum Technologia Alimentaria, 9*(4), 401–412.

32. Potter, N. N., & Hotchkiss, J. H., (2012). *Meat, Poultry, and Eggs, Food Science* (pp. 316–330). Chapter 14. New York: Springer Science & Business Media, Springer Science+Business Media.

33. Pradhan, N., Singh, S., Ojha, N., Shrivastava, A., Barla, A., Rai, V., & Bose, S., (2015). Facets of nanotechnology as seen in food processing, packaging, and preservation industry. *BioMed Research International, 2015*, 1–17. Online; https://doi.org/10.1155/2015/365672.

34. Ramachandraiah, K., Han, S. G., & Chin, K. B., (2015). Nanotechnology in meat processing and packaging: Potential applications: A review. *Asian-Australasian Journal of Animal Sciences, 28*(2), 290–299.

35. Ravichandran, R., (2010). Nanotechnology applications in food and food processing: Innovative green approaches, opportunities and uncertainties for global market. *International Journal of Green Nanotechnology: Physics and Chemistry, 1*(2), P72–P96.

36. Rutzke, C. J., (2003). *Nanoscale Science and Engineering for Agriculture and Food Systems* (p. 21). Report Submitted to the Cooperative State Research, Education and

Extension Service; Washington, DC: The United States Department of Agriculture, National Planning Workshop.

37. Salminen, H., Herrmann, K., & Weiss, J., (2013). Oil-in-water emulsions as a delivery system for n-3 fatty acids in meat products. *Meat Science, 93*(3), 659–667.

38. Salvia-Trujillo, L., Rojas-Graü, M. A., Soliva-Fortuny, R., & Martín-Belloso, O., (2014). Impact of microfluidization or ultrasound processing on the antimicrobial activity against *Escherichia Coli* of lemongrass oil-loaded nanoemulsions. *Food Control, 37*, 292–297.

39. Sanguansri, P., & Augustin, M. A., (2006). Nanoscale materials development-a food industry perspective. *Trends in Food Science & Technology, 17*(10), 547–556.

40. Sharma, B. D., & Sharma, K., (2011). *Outlines of Meat Science and Technology* (p. 106). New Delhi: Jaypee Brothers Medical Publishers (P) Ltd.

41. Silvestre, C., Duraccio, D., & Cimmino, S., (2011). Food packaging based on polymer nanomaterials. *Progress in Polymer Science, 36*(12), 1766–1782.

42. Singh, N., (2015). An overview of prospective application of nanoemulsions in food stuffs and food packaging. *ASIO Journal of Microbiology, Food Science and Biotechnological Innovations (ASIO-JMFSBI), 1*(1), 20–25.

43. Singh, P. K., Jairath, G., & Ahlawat, S. S., (2016). Nanotechnology: A future tool to improve quality and safety in meat industry. *Journal of Food Science and Technology, 53*(4), 1739–1749.

44. Sinha, R., Peters, U., Cross, A. J., Kulldorff, M., Weissfeld, J. L., Pinsky, P. F., Rothman, N., et al., (2005). Meat, meat cooking methods and preservation, and risk for colorectal adenoma. *Cancer Research, 65*(17), 8034–8041.

45. Torab, M., Basti, A. A., & Khanjari, A., (2017). Effect of free and nanoencapsulated forms of *Zataria multiflora*: Essential oil on some microbial and chemical properties of beef burger. *Carpathian Journal of Food Science & Technology, 9*(2), 93–102.

46. U.S. Food and Drug Administration (US FDA), (2017). *Nanotechnology Programs at FDA* (p. 8). https://www.fda.gov/ScienceResearch/SpecialTopics/Nanotechnology/default.htm (accessed on 06 October 2021).

47. Weiss, J., Gibis, M., Schuh, V., & Salminen, H., (2010). Advances in ingredient and processing systems for meat and meat products. *Meat Science, 86*(1), 196–213.

48. Weiss, J., Takhistov, P., & McClements, D. J., (2006). Functional materials in food nanotechnology. *Journal of Food Science, 71*(9), R1–R10.

49. Yang, F. M., Li, H. M., Li, F., Xin, Z. H., Zhao, L. Y., Zheng, Y. H., & Hu, Q. H., (2010). Effect of nano-packing on preservation quality of fresh strawberry (*Fragaria ananassa Duch. cv Fengxiang*) during storage at 4°C. *Journal of Food Science, 75*(3), C236–C240.

50. Yang, M., Kostov, Y., & Rasooly, A., (2008). Carbon nanotubes based optical immuno-detection of staphylococcal enterotoxin B (SEB) in food. *International Journal of Food Microbiology, 127*(1), 78–83.

51. Zhao, X., Yang, Z., Gai, G., & Yang, Y., (2009). Effect of superfine grinding on properties of ginger powder. *Journal of Food Engineering, 91*(2), 217–222.

52. Zhu, K., Huang, S., Peng, W., Qian, H., & Zhou, H., (2010). Effect of ultrafine grinding on hydration and antioxidant properties of wheat bran dietary fiber. *Food Research International, 43*(4), 943–948.

CHAPTER 7

HORIZONS OF NANOTECHNOLOGY IN FOOD SCIENCE

B. MANJULA

ABSTRACT

Nanotechnology has been used as a tool to enhance the texture, nutritional value, and safety of foods. The incorporation of nanostructures into food packaging matrix can improve the barrier properties of matrix and can also be utilized to detect the microorganisms and contaminants. The future of nano-food largely hinges on public perceptions and willingness to accept nano-foods. This chapter reviews applications and issues of nanotechnology in different domains of food.

7.1 INTRODUCTION

The term nanotechnology was coined by Norio Taniguchi for describing controlled thin film deposition [47]. This led to stimulate the scientists to develop protocols and potential tools to perceive structural information of nanomaterials. Nanomaterials can be categorized based on the number dimensions, i.e.: zero-dimensional, one dimensional and two-dimensional materials. For instance, zero dimensional nanoparticles (NPs) have characteristic diameter of ≤ 100 nm, one dimensional NPs have length ≤ 100 nm (nanotubes, nanorods, and nanowires) and two-dimensional nanomaterials have thickness of ≤ 100 nm (nanofilms and nanocoatings) [27, 34]. On the other hand, bulk nanomaterials have characteristic size of ≥ 100 nm (such as: three dimensional nanomaterials). For example, dispersion of nanostructures can be considered as three-dimensional nanomaterials [71].

However, manipulation of the materials at nanoscale influences their bulk properties in comparison to its bulk state [13]. For example, decrease in surface to volume ratio during nanomaterials fabrication results in changing their overall properties and availability of active sites for their functionality [26]. Important nanomaterials properties are: size, shape, charge, solubility, aggregation state and chemical composition [62].

Nanotechnology has assured to improve the food quality and safety through its wide range of principles and processing techniques [67]. However, risks associated with nanotechnology applications in food are still unknown due to unidentified critical limits for foods [65]. The potential advantages of utilizing nanotechnology in food processing include improved food texture; bioavailability of nutrients, flavor, odor, and color. In addition, nanotechnology is also used to improve the existing food products and processes to achieve the desired quality [1, 11, 32, 44]. However, utilization of nanotechnology principles and techniques in food sector is in infancy stage. Nevertheless, the scope for nanotechnology is infinite at all stages of food production (pesticides, post-harvest management) and processing (fortification, packaging, and sensors). Moreover, nano food research groups are concerned more for structuring of food. By keeping in track the current and future needs we can provide various value addition to foods for health and nutrition by decreasing the gauge of interference and the insinuation of structuring of foods. For instance, nanomaterials used in food as preservatives (antimicrobials), fortifiers (passive nanostructures, dried powders) can cross the biological barriers easily compared to its bulk state or macro size. The distribution and persistence of these nanomaterials in human digestive system is unknown and bioaccumulation may lead to potential health hazards on cumulative absorption or adsorption in our body [4, 14, 60]. This chapter gives insights into facets of nanotechnology in different food sectors.

7.2   NANOPHYSICS IN FOOD: SCALAR AND VECTOR QUANTITIES OF FOOD

Nanomaterials are characterized based on their size. However, length is one scale of dimension used for any type of food matrices. In addition, time can be considered in case of passive nanostructures preparation. In an attempt to interact different food structural components, they must come into position at the right time. For example, structuring of nano scale foam that occurs in milliseconds (adsorption of amphiphilic emulsifier at the interface of air-water), whereas subsequent stages of processing occur at the higher length and extended time scale. Addition to this, concepts on soft matter physics are ensuring a new promising way for effectively designing of nano delivery systems for protection of therapeutic nutrients during food processing.

The structure of any food is influenced by different sizes and length scales of its molecules. For instance, liquid foods after crystallization undergo self-assembling, whose size will be only few nm and exhibit higher complexity at the molecular level. Nevertheless, it is still dependent on the underlying principle of "second order phase transition." Although the relation between concepts of soft matter physics and food appears strange at first spectacle, yet all food properties define the soft matter concept [68].

Nano or molecular level control of food properties is essential to achieve the predicted benefit from the food. The relative ratio of biomolecules in the food defines the taste, aroma, and texture of final processed products. Carbohydrates, hydrocolloids (polyelectrolytes), fat, proteins, and water are major biomolecules, which define the nanostructure of food matrices. Hydrocolloids, protein gels, and emulsions are classic examples of soft condensed matter. The crucial contribution of soft matter concepts to nanostructures of food lies in the understanding the relationship among the essential molecules and their functionality in foods and their related physical phenomenon [30]. These concepts interplay with colloidal materials, polymers, surfactants, liquid crystals, and intermolecular forces [30]. An intense influence of soft matter in food system depends on understanding, controlling, and modeling of unit operations across the food manufacturing process. For instance, in spray drying, emulsifier is increased two-folds at the interfacial membranes of nanoemulsion and it evokes the safety concerns of emulsifiers. This information helps in controlling the stability of encapsulated bioactive compound in the food matrices [6].

Modeling of an emulsification system through soft matter concepts has greater impact in optimizing the parameters. For example, microfluidics can be used to mimic the different flow regimes in the homogenizer. In this process, nanoemulsion droplets can be prepared with different amounts of adsorbed emulsifier at the interface and with different forces [35]. However, low-surface active materials adsorbed droplets may collide at high impact; and as adsorption at interface increases, they collide at low impact in the homogenizer. These results can be utilized to design a homogenizer flow profile that is optimized for the type of nanoemulsion to be processed [35].

A rational, effective way to design and evaluate the available options for an encapsulation system is with the help of a scientific retro-design approach [68]. However, the primary objective of designing an encapsulation system is often restricted to the transport of functional compounds into targeted food matrices. Therefore, there is a need to concentrate on improving the characteristic properties of an encapsulation system in such a way so that it enhances the functionality and stability of functional compounds in a compatible food matrix. Hence being crucial in designing encapsulation vehicles, the concept of soft matter science is important to understand the performance of functional molecules in food.

7.3 PERCEPTIONS OF NANOTECHNOLOGY IN DIFFERENT FOOD SECTORS

Foods are naturally composed of complex nanostructure assemblies. The chemical and physical properties of these assemblies will determine the stability of dissolved colloidal mixtures (lipids, proteins, additives, and carbohydrates) in the food matrices [10, 48, 54, 56]. In addition, manipulation of nanostructures at nano scale provides the structuring of food matrices with added value and functionality [23]. Nanotechnology holds forth remarkable clutches not only in designing or structuring food matrices; but also, for food environment. For instance, nanotechnology cannot be restricted to structure the food matrices using different nano ingredients, but it can also be exploited to develop new type of food packaging materials, sensors, and smart labeling [14]. However, nanoscale research in different food sectors is mainly focused on shelf-life extension of product using nano-edible coatings [15]. In addition, nanotechnology-based encapsulation can significantly prolong the functional molecule residence time in the gastrointestinal tract by reducing the intestinal clearance (via faster absorption) and increasing the surface interactions with the host cells or materials [31, 58].

In recent years, nanotechnology is a controversial subject in concern with food encapsulation and its safety [4, 17]. Hence focusing more on nanotechnology-based sensors, packaging materials and labeling may be a stepping stone for food industrialists and can help the general public to know benefits of nanotechnology in the food sector. Nanotechnology in packaging has become the integral part of intelligent and active packaging with improved barrier properties [60, 66]. In active packaging, nanotechnology plays a passive role in extending the shelf-life of a food product with improved barrier properties and eventually ensuring the safety and quality. In case of intelligent packaging, nanotechnology-based sensors can help in indicating the quality and safety (microbial quality, chemical deterioration) of food products during storage.

In a novel application, nanotechnology can also be used to manipulate the functional properties of enzyme to deliver health and nutrition benefits. For instance, to increase the bioavailability of essential micro-nutrients (such as: vitamins, minerals), it is often suggested to hydrolyze the anti-nutritional factors with the help of enzymes [70]. In this regard, increasing the active sites in the enzymes using nanotechnology can help in reducing the minimum enzyme concentration required in the food [8]. In addition, using nanomaterials for enhancing the functionality of enzymes will be cost-effective and highly active due to their high surface to volume ratio compared to its macroscale structure [33, 46, 53].

7.3.1 NANOTECHNOLOGY IN MEAT PROCESSING

The ingredients and additives used for enhancing the quality and safety of muscle foods are increasing. Antioxidants (BHT (butylated hydroxy toluene), BHA (butylated hydroxyl anisole), and tocopherols), water binders (sodium caseinate), thickeners, humectants, curing agents (sodium nitrite, sodium nitrate and sodium erythrobate), flavor enhancer (monosodium glutamate), enzymes (papain, ficin, and bromelin) and sweeteners are commonly used additives in the meat processing industry [18, 59].

In meat processing, fat replacers (use of citrus fiber, oat fiber, soy protein concentrate), fat profile modification (use of flaxseed or linseed oil extracts), salt reduction (use of apple pulp, seaweeds), nitrite reduction (use of spinach and celery juice) and delivery of antioxidants (use of rosemary extract) are few examples of nanotechnology applications in meat processing [19, 63, 64]. In addition, use of nano scale powdered ingredients instead of bulk ingredients helps in enhancing their activity. For example, decreasing the particle size of tea enhances the antioxidant activity [2].

In meat processing, ginger is used as meat tenderizer and extender. Changing the structural morphology of ginger into powder and reduction in particle size will improve its penetrability and solubility [74]. At packaging level, silver nanoparticles (AgNPs) are used to improve the safety of meat foods. In an *in-situ* study, AgNPs reduced the microbial level in beef meet under modified atmospheric storage [20].

7.3.2 NANOTECHNOLOGY IN BAKERY PRODUCTS

Nanotechnology presents a promising future to fortify the bakery food matrices with different delivery systems. The release of nutrients at a controlled manner during processing is of interest in bakery processing. For example, after baking of product, if encapsulated compound needs to be available for human consumption (nutrients), then nanotechnology-based delivery systems can be used [21, 72]. If shelf-life is concerned, encapsulated antimicrobials can be used; however minimum inhibitory concentration should be established.

Encapsulation to deliver the flavors for enhanced eating experience in baked foods is of interest in bakery industry. However, nanoencapsulation can be ingredient specific and environmental condition sensitive for baked food matrices. The ingredients used in bakery food preparations can interact with encapsulating nanostructures and may trigger the release of functional molecules. In addition, during baking process, hardening of food matrix can restrict the movement of encapsulated nanostructure inside the food compared to dough matrix, which results in cluster formation followed by non-uniform sensory feel in different parts of the food.

Hence, it is essential to ensure homogeneous distribution of passive nanostructures (delivery systems or encapsulation matrices) in the baking premixes or during dough formation. In general, nanotechnology-based encapsulation can be used in bakery processing principally at the ingredient level. For example, using encapsulated dough conditioners, anti-stalling agents, antimicrobials, leavening agents, sweeteners, flavors, and nutrients can enhance the quality and safety aspects of baked foods without interfering with sensory attributes (Figure 7.1).

Nanotechnology based functional confectioneries are currently progressing in market. Viative (vitamin and calcium containing soft chewy candy), Orbit, mentholated lozenges are few successful examples of confectionery products that were developed with nanotechnology (encapsulation). Regardless of these developments, commercial success is hindered for therapeutic healthy confectionary foods due to complex structural organizations of food matrices. In case of flavor encapsulation and delivery for baked foods, the sole concern is solubilizing the functional compound and controlled release over time in response to surrounding environment in gastrointestinal tract. In addition, delivery of functional therapeutic molecules necessitates additional intricate design and understanding their biological fate (adsorption, distribution, metabolism, and elimination) across the membranes.

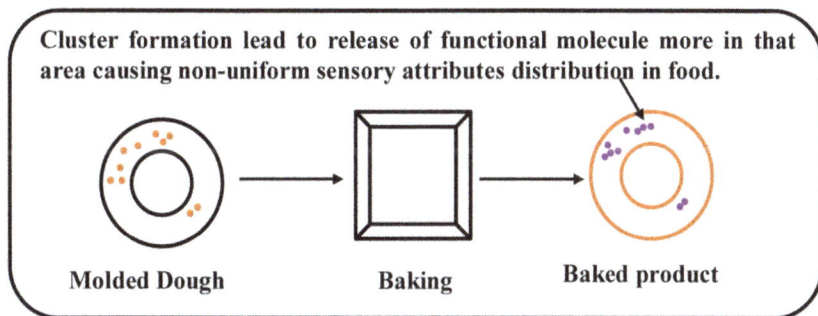

Cluster formation lead to release of functional molecule more in that area causing non-uniform sensory attributes distribution in food.

Molded Dough   **Baking**   **Baked product**

FIGURE 7.1   Importance of encapsulated passive nanostructures distribution in the bakery matrix.

7.3.3 NANOTECHNOLOGY IN DAIRY PROCESSING

Proteins (casein, lactoglobulin) in milk are in nano scale range. Passive nanostructures (such as: nanoemulsions, liposomes, NPs, and nanotubes) have been designed using milk nanostructures [7, 29]. Nanotechnology has bright future in dairy processing, starting from feeding of animals, milk processing and packaging to overcome the safety and quality issues. Nano-filtration, nanocoating for dairy equipment and products, nano-preservatives, and effluent treatment are few such examples. The foremost application of nanotechnology in dairy industry is encapsulation of bioactive compounds for improving nutritional status [16]. Dairy based products are highly nutritious and exhibit lower shelf-life. However, the deterioration kinetics is high for dairy products compared to grains and vegetables [3, 9, 12].

The public awareness of nanotechnology application in dairy industry is significant because of their alleged negative effects on health and environment. The nanomaterials used for fortification, encapsulation or packaging must be eco-friendly. Therefore, to reach the nutritional security in dairy, nanotechnology is a viable option and further studies are vital to examine the threats of nanomaterials on human health [24].

7.4   NANOTECHNOLOGY VERSUS FUNCTIONAL FOODS

Number of key nutrients in food products are low compared to their unprocessed status, either due to cultural differences (high salt, fat, and sugar) or processing methods [57]. For a healthy life style, key nutrients (vitamins, minerals, and essential fatty acids) play major role. To enhance the nutritional quality in dairy supply chain, nanotechnology intervention is crucial according to many researchers while sustaining low cost [41, 46, 55, 63]. Currently, food industry is looking for use of nanotechnology to design the functional food matrices with the aid of nano delivery mechanisms (fortification, enrichment, and encapsulation methods) for essential nutrients and to enhance their efficacies in the food. The fabrication of effective nanoparticle-based delivery systems for functional foods with enhanced therapeutic activity requires deeper knowledge of biological fate of individual nutrients at their nanoscale range [22, 51].

The development of functional foods is more concomitant with incorporation of bioactive molecules [57]. Compared to direct incorporation, encapsulated form of bioactive molecules in any matrix results in better utilization of functionality of bioactive compounds related to their controlled release.

7.4.1   DELIVERY SYSTEMS

Reducing the size of functional compounds to nanoscale was well-demonstrated method to enhance their absorption in the digestion tract. The functional molecules can be extracted and then introduced in a shell matrix (delivery system), which can be efficiently and instantly absorbed by the cell membranes in our body. Currently, several natural and synthetic shell matrices are available for design of encapsulation matrices or delivery systems [28, 39].

Delivery systems are passive nanostructures to protect the labile bioactive compounds in the food during processing. In general, delivery systems can be classified as shell and core-based encapsulation matrices that can be used for both lipophilic and hydrophilic functional molecules. In addition, these delivery systems with nano scale range possess distinct properties, such as: size, shape, charge, and other surface properties [43]. They have special advantages compared to macro-scale encapsulating matrices, for instance: targeted delivery, stimuli-based release (pH, enzyme action), site-specific, and controlled release of encapsulated functional molecules [37, 38]. However, utilizing these passive nanostructures in food can change the functionality of existing matrix, such as: improved texture, color of the products, reactivity of special ingredients and consistency of matrix [25, 43, 69, 72]. Nevertheless, increased surface to volume ratio will enhance the absorption rate of an encapsulated compound in gastrointestinal tract.

Bioactive compounds are synonymously used as nutraceuticals [17, 36, 42]. In addition to naturally existing bioactive compounds (vitamins, peptides, and antioxidants), food delivery systems are used for enriching and fortifying with these functional compounds, and which also includes probiotics, prebiotics, and other molecules associated with therapeutic potential [17, 45, 50].

The bioavailability and bioaccessibility of these functional molecules depend on their physiological action in the digestive tract, their stability during processing and food structures [43]. For instance, protein-based delivery systems denature at higher temperature; molecules with more functional groups in delivery systems increases the molecular interactions [43, 73]. Nevertheless, most important challenges endure the understanding of the fraction of bioactive compounds or nutrients that can be released from the food matrix and are consequently potentially accessible for absorption within gastrointestinal tract [40, 43, 49].

Nutraceuticals deliver nutrients that have health benefits or therapeutic value and through nanoencapsulation using delivery vehicles will enhance their efficacy and bioavailability in food. Regulations are perhaps the major barrier for commercialization of this technology. Hence, governments should provide regulatory landscape framework [52].

7.5 SUMMARY

This chapter focuses on facets of nanotechnology in different food sectors. Encapsulation, nanocomposite development, functional foods, NPs-based delivery systems are the successful nanotechnology interventions in food sectors. Perspectives of nanotechnology in different food sectors (such as: meat, bakery, confectionery, and dairy processing) have been briefly discussed.

KEYWORDS

- bioavailability
- biomolecules
- nanoemulsions
- nanoparticles
- nanotechnology
- nutraceuticals
- toxicity

REFERENCES

1. Aider, M., (2010). Chitosan application for active bio-based films production and potential in the food industry: Review. *LWT-Food Science and Technology, 43*(6), 837–842.
2. Astill, C., Birch, M. R., Dacombe, C., Humphrey, P. G., & Martin, P. T., (2001). Factors affecting the caffeine and polyphenol contents of black and green tea infusions. *Journal of Agricultural and Food Chemistry, 49*(11), 5340–5347.
3. Birwal, P., Rangi, P., & Ravindra, M. R., (2019). Nanotechnology applications in packaging of dairy and meat products. Chapter 1. In: Lohith, K. D. H., Goyal, M. R., & Suleria, H. A. R., (eds.), *Nanotechnology Applications in Dairy Science: Packaging, Processing, Preservation* (pp. 3–26). Oakville - ON: Apple Academic Press.
4. Bouwmeester, H., Dekkers, S., Noordam, M. Y., Hagens, W. I., Bulder, A. S., De Heer, C., Ten, V. S. E., Wijnhoven, S. W., et al., (2009). Review of health safety aspects of nanotechnologies in food production. *Regulatory Toxicology and Pharmacology, 53*(1), 52–62.
5. Bradley, E. L., Castle, L., & Chaudhry, Q., (2011). Applications of nanomaterials in food packaging with a consideration of opportunities for developing countries. *Trends in Food Science & Technology, 22*(11), 604–610.
6. Cambiella, A., Benito, J. M., Pazos, C., & Coca, J., (2007). Interfacial properties of oil-in-water emulsions designed to be used as metalworking fluids. *Colloids and Surfaces A: Physicochemical and Engineering Aspects, 305*(1), 112–119.

7. Carr, A., & Golding, M., (2016). Functional milk proteins production and utilization: Casein-based ingredients. In: *Advanced Dairy Chemistry* (pp. 35–66). Cham: Springer.

8. Cushen, M., Kerry, J., Morris, M., Cruz-Romero, M., & Cummins, E., (2012). Nanotechnologies in the food industry-recent developments, risks and regulation. *Trends in Food Science & Technology, 24*(1), 30–46.

9. Dasarahalli-Huligowda, L. K., Goyal, M. R., & Suleria, H. A. R., (2019). *Nanotechnology Applications in Dairy Science: Packaging, Processing, and Preservation* (1ˢᵗ edn., p. 275). Oakville - ON: Apple Academic Press.

10. De Azeredo, H. M., (2013). Antimicrobial nanostructures in food packaging. *Trends in Food Science & Technology, 30*(1), 56–69.

11. De Moura, M. R., Mattoso, L. H., & Zucolotto, V., (2012). Development of cellulose-based bactericidal nanocomposites containing silver nanoparticles and their use as active food packaging. *Journal of Food Engineering, 109*(3), 520–524.

12. DH Kumar, L., & Sarkar, P., (2018). Potential of nanotechnology in dairy processing: A review. Chapter 3. In: Megh, R. G., (ed.), *Sustainable Biological Systems for Agriculture: Emerging Issues in Nanotechnology, Biofertilizers, Wastewater, Farm Machines* (pp. 55–80). Oakville - ON: Apple Academic Press.

13. Duan, H., Wang, D., & Li, Y., (2015). Green chemistry for nanoparticle synthesis. *Chemical Society Reviews, 44*(16), 5778–5792.

14. Duncan, T. V., (2011). Applications of nanotechnology in food packaging and food safety: Barrier materials, antimicrobials and sensors. *Journal of Colloid and Interface Science, 363*(1), 1–24.

15. Duran, N., & Marcato, P. D., (2013). Nanobiotechnology perspectives. Role of nanotechnology in the food industry: A review. *International Journal of Food Science & Technology, 48*(6), 1127–1134.

16. Erfanian, A., Rasti, B., & Manap, Y., (2017). Comparing the calcium bioavailability from two types of nano-sized enriched milk using *in-vivo* assay. *Food Chemistry, 214*, 606–613.

17. Ezhilarasi, P., Karthik, P., & Chhanwal, N., (2013). Nanoencapsulation techniques for food bioactive components: A review. *Food and Bioprocess Technology, 6*(3), 628–647.

18. Falowo, A. B., Fayemi, P. O., & Muchenje, V., (2014). Natural antioxidants against lipid-protein oxidative deterioration in meat and meat products: A review. *Food Research International, 64*, 171–181.

19. Fernández-Ginés, J. M., Fernández-López, J., Sayas-Barberá, E., & Pérez-Alvarez, J., (2005). Meat products as functional foods: A review. *Journal of Food Science, 70*(2), R37–R43.

20. Fernández, A., Picouet, P., & Lloret, E., (2010). Reduction of the spoilage-related microflora in absorbent pads by silver nanotechnology during modified atmosphere packaging of beef meat. *Journal of Food Protection, 73*(12), 2263–2269.

21. Ganesh, V., & Hettiarachchy, N. S., (2016). A review: Supplementation of foods with essential fatty acids-can it turn a breeze without further ado? *Critical Reviews in Food Science and Nutrition, 56*(9), 1417–1427.

22. Giles, E. L., Kuznesof, S., Clark, B., Hubbard, C., & Frewer, L. J., (2015). consumer acceptance of and willingness to pay for food nanotechnology: A systematic review. *Journal of Nanoparticle Research, 17*(12), 1–26.

23. Graveland-Bikker, J., & De Kruif, C., (2006). Unique milk protein based nanotubes: Food and nanotechnology meet. *Trends in Food Science & Technology, 17*(5), 196–203.

24. Gruère, G. P., (2012). Implications of nanotechnology growth in food and agriculture in OECD countries. *Food Policy, 37*(2), 191–198.

25. Guichard, E., (2002). Interactions between flavor compounds and food ingredients and their influence on flavor perception. *Food Reviews International, 18*(1), 49–70.

26. Hannon, J. C., Kerry, J., Cruz-Romero, M., Morris, M., & Cummins, E., (2015). Advances and challenges for the use of engineered nanoparticles in food contact materials. *Trends in Food Science & Technology, 43*(1), 43–62.

27. Hartings, M. R., Benjamin, N., Briere, F., Briscione, M., Choudary, O., Fisher, T. L., Flynn, L., et al., (2016). Concurrent zero-dimensional and one-dimensional biomineralization of gold from a solution of Au^{3+} and bovine serum albumin. *Science and Technology of Advanced Materials, 14*(6), 1–8.

28. Hati, S., Makwana, M. R., & Mandal, S., (2019). Encapsulation of probiotics for enhancing the survival in gastrointestinal tract. Chapter 9. In: Lohith, K. D. H., Goyal, M. R., & Suleria, H. A. R., (eds.), *Nanotechnology Applications in Dairy Science: Packaging, Processing, Preservation* (pp. 225–244). Oakville - ON: Apple Academic Press.

29. Hayashi, M., Silanikove, N., Chang, X., Ravi, R., Pham, V., Baia, G., Paz, K., et al., (2015). Milk derived colloid as a novel drug delivery carrier for breast cancer. *Cancer Biology & Therapy, 16*(8), 1184–1193.

30. Hirst, L. S., (2012). *Fundamentals of Soft Matter Science* (p. 219). Boca Raton, FL: CRC Press.

31. Huang, Q., Yu, H., & Ru, Q., (2010). Bioavailability and delivery of nutraceuticals using nanotechnology. *Journal of Food Science, 75*(1), R50–R57.

32. Hyldgaard, M., Mygind, T., & Meyer, R. L., (2012). Essential oils in food preservation: Mode of action, synergies, and interactions with food matrix components. *Frontiers in Microbiology, 3*, 12–19.

33. Jia, H., Zhu, G., & Wang, P., (2003). Catalytic behaviors of enzymes attached to nanoparticles: The effect of particle mobility. *Biotechnology and Bioengineering, 84*(4), 406–414.

34. Kim, P. Y., Oh, J. W., & Nam, J. M., (2015). Controlled co-assembly of nanoparticles and polymer into ultralong and continuous one-dimensional nanochains. *Journal of the American Chemical Society, 137*(25), 8030–8033.

35. Krebs, T., Schroen, K., & Boom, R., (2012). Microfluidic method to study demulsification kinetics. *Lab on a Chip, 12*(6), 1060–1070.

36. Kumar, D. D., Mann, B., Pothuraju, R., Sharma, R., & Bajaj, R., (2016). Formulation and characterization of nanoencapsulated curcumin using sodium caseinate and its incorporation in ice cream. *Food & Function, 7*(1), 417–424.

37. Kumar, D. L., Mitra, J., & Roopa, S., (2020). Nanoencapsulation of food carotenoids. Chapter 7. In: Rajan, S., Nandita, D., & Eric, L., (eds.), *Environmental Nanotechnology* (Vol. 3, pp. 203–242). Cham: Springer.

38. Kumar, D. L., & Sarkar, P., (2017). Nanoemulsions for nutrient delivery in food. Chapter 4. In: Ranjan, S., Dasgupta, N., & Lichtfouse, E., (eds.), *Nanoscience in Food and Agriculture* (Vol. 5, pp. 81–121). Cham: Springer.

39. Lohith, K. D. H., & Sarkar, P., (2018). Encapsulation of bioactive compounds using nanoemulsions. *Environmental Chemistry Letters, 16*(1), 59–70.

40. Mao, Y., & McClements, D. J., (2012). Fabrication of functional micro-clusters by heteroaggregation of oppositely charged protein-coated lipid droplets. *Food Hydrocolloids, 27*(1), 80–90.

41. McClements, D. J., (2015). Reduced-fat foods: The complex science of developing diet-based strategies for tackling overweight and obesity. *Advances in Nutrition: An International Review Journal, 6*(3), 338S–352S.

42. McClements, D. J., Decker, E. A., & Weiss, J., (2007). Emulsion-based delivery systems for lipophilic bioactive components. *Journal of Food Science, 72*(8), R109–R124.

43. McClements, D. J., & Rao, J., (2011). Food-grade nanoemulsions: Formulation, fabrication, properties, performance, biological fate, and potential toxicity. *Critical Reviews in Food Science and Nutrition, 51*(4), 285–330.

44. Mohammed, F. A., Balaji, K., Girilal, M., Kalaichelvan, P., & Venkatesan, R., (2009). Mycobased synthesis of silver nanoparticles and their incorporation into sodium alginate films for vegetable and fruit preservation. *Journal of Agricultural and Food Chemistry, 57*(14), 6246–6252.

45. Mohanty, D., Jena, R., Choudhury, P. K., Pattnaik, R., Mohapatra, S., & Saini, M. R., (2016). Milk derived antimicrobial bioactive peptides: A review. *International Journal of Food Properties, 19*(4), 837–846.

46. Motornov, M., Zhou, J., Pita, M., Tokarev, I., Gopishetty, V., Katz, E., & Minko, S., (2009). An integrated multifunctional nanosystem from command nanoparticles and enzymes. *Small, 5*(7), 817–820.

47. Mulvaney, P., (2015). Nanoscience versus nanotechnology, defining the field. *ACS Nano, 9*(3), 2215–2217.

48. Neethirajan, S., & Jayas, D. S., (2011). Nanotechnology for the food and bioprocessing industries. *Food and Bioprocess Technology, 4*(1), 39–47.

49. Noshad, M., Mohebbi, M., Shahidi, F., & Koocheki, A., (2015). Effect of layer-by-layer polyelectrolyte method on encapsulation of vanillin. *International Journal of Biological Macromolecules, 81*, 803–808.

50. Pan, Y., & Nitin, N., (2015). Effect of layer-by-layer coatings and localization of antioxidant on oxidative stability of a model encapsulated bioactive compound in oil-in-water emulsions. *Colloids and Surfaces B: Biointerfaces, 135*, 472–480.

51. Raman, M., & Doble, M., (2019). Polyphenol nanoformulations for cancer therapy: Role of milk components. Chapter 6. In: Lohith K. D. H., Goyal, M. R., & Suleria, H. A. R., (eds.), *Nanotechnology Applications in Dairy Science: Packaging, Processing, Preservation* (pp. 123–168). Oakville - ON: Apple Academic Press.

52. Ramkumar, C., Vishwanatha, A., & Saini, R., (2019). Regulatory aspects of nanotechnology for food industry. Chapter 7. In: Lohith K. D. H., Goyal, M. R., & Suleria, H. A. R., (eds.), *Nanotechnology Applications in Dairy Science: Packaging, Processing, Preservation* (pp. 169–184). Oakville - ON: Apple Academic Press.

53. Rao, J., & McClements, D. J., (2013). Optimization of lipid nanoparticle formation for beverage applications: Influence of oil type, cosolvents, and cosurfactants on nanoemulsion properties. *Journal of Food Engineering, 118*(2), 198–204.

54. Raynes, J. K., Carver, J. A., Gras, S. L., & Gerrard, J. A., (2014). Protein nanostructures in food-should we be worried? *Trends in Food Science & Technology, 37*(1), 42–50.

55. Saberi, A. H., Fang, Y., & McClements, D. J., (2013). Fabrication of vitamin E-enriched nanoemulsions: Factors affecting particle size using spontaneous emulsification. *Journal of Colloid and Interface Science, 391*, 95–102.
56. Sanguansri, P., & Augustin, M. A., (2006). Nanoscale materials development - a food industry perspective. *Trends in Food Science & Technology, 17*(10), 547–556.
57. Sarkar, P., Lohith, K. D. H., Dhumal, C., Panigrahi, S. S., & Choudhary, R., (2015). Traditional and ayurvedic foods of Indian origin. *Journal of Ethnic Foods, 2*(3), 97–109.
58. Scrinis, G., & Lyons, K., (2007). The emerging nano-corporate paradigm: Nanotechnology and the transformation of nature, food and agri-food systems. *International Journal of Sociology of Agriculture and Food, 15*(2), 22–44.
59. Shah, M. A., Bosco, S. J. D., & Mir, S. A., (2014). Plant extracts as natural antioxidants in meat and meat products. *Meat Science, 98*(1), 21–33.
60. Siegrist, M., Stampfli, N., Kastenholz, H., & Keller, C., (2008). Perceived risks and perceived benefits of different nanotechnology foods and nanotechnology food packaging. *Appetite, 51*(2), 283–290.
61. Silvestre, C., Duraccio, D., & Cimmino, S., (2011). Food packaging based on polymer nanomaterials. *Progress in Polymer Science, 36*(12), 1766–1782.
62. Smolkova, B., El Yamani, N., Collins, A. R., Gutleb, A. C., & Dusinska, M., (2015). Nanoparticles in food. Epigenetic changes induced by nanomaterials and possible impact on health. *Food and Chemical Toxicology, 77*, 64–73.
63. Sowbhagya, H. B., (2014). Chemistry, technology, and nutraceutical functions of celery (*Apium graveolens* L.): An overview. *Critical Reviews in Food Science and Nutrition, 54*(3), 389–398.
64. Sowbhagya, H. B., Lohith, K. D. H., Anush, S. M., & Mohan, R. L., (2015). Microwave impact on the flavor compounds of cinnamon bark (*Cinnamomum Cassia*) volatile oil and polyphenol extraction. *Current Microwave Chemistry, 2*, 115–122.
65. Sozer, N., & Kokini, J. L., (2009). Nanotechnology and its applications in the food sector. *Trends in Biotechnology, 27*(2), 82–89.
66. Stampfli, N., Siegrist, M., & Kastenholz, H., (2010). Acceptance of nanotechnology in food and food packaging: A path model analysis. *Journal of Risk Research, 13*(3), 353–365.
67. Tiede, K., Boxall, A. B., Tear, S. P., Lewis, J., David, H., & Hassellöv, M., (2008). Detection and characterization of engineered nanoparticles in food and the environment. *Food Additives and Contaminants, 25*(7), 795–821.
68. Ubbink, J., (2012). Soft matter approaches to structured foods: From "cook-and-look" to rational food design. *Faraday Discussions, 158*(1), 9–35.
69. Ubbink, J., & Krüger, J., (2006). Physical approaches for the delivery of active ingredients in foods. *Trends in Food Science & Technology, 17*(5), 244–254.
70. Valdés, M. G., González, A. C. V., Calzón, J. A. G., & Díaz-García, M. E., (2009). Analytical nanotechnology for food analysis. *Microchimica Acta, 166*(1–2), 1–19.
71. Wan, Y., Yang, Z., Xiong, G., Guo, R., Liu, Z., & Luo, H., (2015). Anchoring Fe_3O_4 nanoparticles on three-dimensional carbon nanofibers toward flexible high-performance anodes for lithium-ion batteries. *Journal of Power Sources, 294*, 414–419.

72. Wang, T., Soyama, S., & Luo, Y., (2016). Development of a novel functional drink from all natural ingredients using nanotechnology. *LWT-Food Science and Technology, 73,* 458–466.

73. Wei, Z., & Gao, Y., (2016). Physicochemical properties of β-carotene bilayer emulsions coated by milk proteins and chitosan-EGCG conjugates. *Food Hydrocolloids, 52,* 590–599.

74. Zhao, X., Yang, Z., Gai, G., & Yang, Y., (2009). Effect of superfine grinding on properties of ginger powder. *Journal of Food Engineering, 91*(2), 217–222.

CHAPTER 8

ROLE OF NANOTECHNOLOGY IN SELECTIVE TARGETING OF CANCER

RAHUL SAINI

ABSTRACT

Nanoparticles (NPs) have shown potential to overcome several limitations in conventional drug delivery methods. Non-selective targeting, reduced drug localization, less-cellular uptake rate are issues associated with conventional chemotherapeutic therapies. NPs can be programed for cancer-cell recognition to get direct access to cancerous cell avoiding the interaction with healthy cells. Combining NPs with different biological systems (such as: ligand, signaling molecules specific to receptor) will not only enhance the uptake of cellular rate but also increase the drug localization to a specific site. Different types of interactions during the drug entry into the body, difference between nanotechnology and conventional cancer targeting methods, and ethics associated with use of NPs are also discussed.

8.1 INTRODUCTION

The cancer results from genetic mutation, hence requires complex treatment processes [3]. Variety of treatment processes are available along with side-effects and limitations [62]. For instance, surgical removal of cancer mass, radiation, hormone therapy and chemotherapy are widely known cancer treatment methods.

Chemotherapy involves use of anti-cancer drug and has the ability to quench the cancer cell proliferation [18]. However, the challenges associated with the chemotherapy are non-specific; and inefficient drug delivery system causes the induction of side-effects of anti-cancer drugs. Major problem in which all other methods also fail is the inability to distinguish between normal cells and cancer cells, which influence the working of normal human organs [38]. It is the time during which drug should be engineered in such as way so that they should not lose their activity and must have site directed effect of target cells or organ.

Nanotechnology has revolutionized the medicinal drug delivery system [34, 52]. Nanotechnology involves the use two approaches: (a) bottom-up approach: Building up of nanostructures from atoms and closely intimating the chemical structure; and (b) top-down approach: Reducing the large sized structures to their smallest form, such as, applications of photonics in nano-engineering and nano-electronics [41]. Nanotechnology has come up as a promising technique for treatment and diagnosis of cancer at molecular level [53]. Number of research studies have been in progress for development of nano-based cancer treatments along with reduced side-effects. It has also the capability to assist the drug to cross the blood brain barrier.

Currently some of the examples of nanotechnology-based drug delivery system are: dendrimers, liposomes, nanocapsules, nanosphere, carbon nanotubes (CNTs) [42, 48]. Some of commercial examples are Abraxane and Doxil [39]. Nanomaterials help to bridge the gap between molecules and atoms. Major properties are small size, high surface area, and ability to tag with a biological molecule. In general, biological molecules in use are collagen or antibodies. Core of nano-biomaterial is usually formed by nanomaterials. This core is then surrounded by the inert and biocompatible material, such as, silica. For further surface optimization, linkers were linked to the surface that is often able to link the biomolecules to the surface depending on requirement [21]. This chapter is focused on recent development of nanotechnology in drug delivery systems.

8.2   LIMITATIONS OF CHEMOTHERAPY

Chemotherapy agents actively attack the rapidly dividing cells, which is the characteristic property of cancer cells. This is the major reason that these agents also hinder the normal body cells that generally have high dividing rate, such as, digestive tract, macrophages, bone marrow and hair follicles [62]. The major anomaly of conventional chemotherapy techniques is non-selective action resulting in severe side-effects of using chemotherapy agents. For instance, the side effects include: mucositis, myelosuppression, organ dysfunction, anemia, and alopecia. These may cause treatment delay, dosage variation and therapy discontinuance [8].

Cells surrounding the solid tumors are generally ceased by the action of chemotherapy along with death of tumor cells. Furthermore, agents are sometime not able to penetrate the tumor cells till their core, failing to eradicate the cancerous cells [55].

Chemotherapeutic agents are generally removed from the circulation by action of macrophages. Hence, it is not able to keep contact with cancer cells for longer period of time. Another problem associated with these agents is the poor solubility making them unable to penetrate through the blood brain barrier [38]. P-glycoprotein (a multidrug resistance protein) is often overexpressed on the cancerous cell; and prevents the accumulation of drugs by fluxing the out the drug molecules and letting the development of resistance to anticancer drugs [25, 53]. Thus, it is the time available for development of safe, accurate, fast, and cost-effective method for targeting of cancer.

8.3 TYPES OF CANCER TARGETING

8.3.1 ACTIVE TARGETING

In active targeting method, nanoparticles (NPs) comprising the chemo-therapeutic agents are directly interacting with the cancerous cells. Its interaction is based on molecular recognition technique. Usually, targeting agents are attached with the surface of NPs for molecular recognition. NPs interacts with the neoplastic cells through antibody-antigen interaction or ligand-receptor recognition [6, 16, 40]. Nanotechnology system have three system: (i) anti-cancer drug; (ii) a targeting moiety-penetration enhancer; and (iii) a carrier. Materials generally employed for construction of nanoparticle are: lipids, metals, polymers, and ceramic [60]. Synthetic or natural lipids and polymers are commonly utilized as vectors [11, 13, 28].

In general, chemotherapeutic particles are removed by macrophages through the process of phagocytosis. The variety of strategies are available to increase the availability of NPs in blood circulation. Coating of NPs with hydrophilic polymers helps to evade wash out and increase their bioavailability in the blood stream. This is known as cloud effect

[53, 54]. Some of the examples of hydrophilic polymer are: polyamines, polysaccharides, poloxamers, and polyethylene glycol [51, 57]. There are various unique properties of cancer cells at molecular level that result in their differentiation from normal cells. Nanoparticle binds with the overexpressed receptor on the cell surface and releases the loaded drug inside the cells. Study was conducted to investigate the ligand-receptor interaction for clinical use [45].

8.3.2   PASSIVE TARGETING

As apoptosis is stopped in cancerous cells, they continue sucking nutritious agents abnormally through the blood vessels forming wide and leaky blood vessels around the cells induced by angiogenesis. The size of the pores in leaky endothelial cells ranges from 100 to 780 nm. Thus, NPs below this size can easily pass through the pores [19]. As a result, it facilitates to efflux the NPs to cluster around the neoplastic cells.

NPs can be targeted to specific area of capillary endothelium, to concentrate the drug within a particular organ and perforate the tumor cells by passive diffusion or convection. Lack of lymphatic drainage eases the diffusion process. Drugs that enter the interstitial area may extend the retention time. This feature is called the enhanced permeability and retention (EPR) effect and it facilitates tumor interstitial drug accumulation [33]. NPs can easily accumulate selectively by EPR effect and then diffuse into the cells [53].

8.4   BIOSYNTHESIS OF NANO-DRUG DELIVERY VEHICLES

Synthesis of nanocarriers plays important role in development of novel NPs in cancer targeting and diagnosis. Furthermore, surface area, high inherent area, and particle stability are valuable properties in drug delivery and medical applications. Various metal NPs have been utilized as drug carriers due to their properties and nanostructures. For instance, Malathi et al. [35] reported the gold nanoparticle synthesis for drug delivery based on chitosan as the capping agent. They employed the oil-in-water emulsions for rifampicin release in the body. While, silver nanoparticle was prepared using egg whites and silver nitrate. This resulted in biomolecules loaded silver nanoparticle having spherical shapes with size of 20 nm. Silver nanoparticle showed the biocompatibility with mouse fibroblast cells during cancer radiation therapy [32].

Seaweeds also have promising role in preparation of nanometals. For instance, gold nanoparticle from gold chloride was prepared in trivalent aurum solutions along with leaf extracts of brown seaweed (*Padina gymospora*) [50]. Huge work is in progress for development of carriers specifically for liver cancer targeting. Pullulan-supported gold NPs were linked with 5-fluorouracil and FA to produce new targeted drug imaging and delivery. The *in-vitro* cytotoxicity of free Pullulan gold NPs on tumor cells showed that the concentration of free Pullulan gold NPs to achieve 50% of growth of inhibition was much lower [14]. To increase the bioavailability and sustained release of drug molecule, polyphenol of strawberry extracts linked with positively protonated amino groups of chitosan were utilized. It also helped to achieve maximum encapsulation [49].

Curcumin and emu oil from emu birds have shown promising results against inflammation. The nanoemulsions were prepared using Cremophor, Labrafil, and emu oil. It showed good anti-inflammatory activity when mixed with curcumin [20]. For preparation of silver nanoparticles (AgNPs), the inexpensive, fast, and renewable strategy was performed in aqueous media utilizing light emitted as the catalyst. The prepared nanocomposite exhibited very high antimicrobial activity against various strains of microorganisms. Another green and clean strategy for the AgNPs preparation used the *Delftia*'s cell free suspension and demonstrated the both drug delivery agents and antifungal activities [26]. Ginsenoside NPs play important role in carrying insoluble drugs. Dai et al. [9] demonstrated that anticancer drug has enhanced *in-vivo* and *in-vitro* effects when loaded on ginsenoside NPs than the free drug.

8.5   ROLE OF TARGETING AGENT

NPs or nanocarriers (such as: dendrimers, liposomes, micelles, nanospheres, and carbon nanotube) loaded with anticancer drugs, polymers or therapeutic protein are used as targeting agents for cancer treatment [27, 31]. Drug molecules can be covalently attached, entrapped, adsorbed, or encapsulated to the nanocarriers [45, 48]. Liposomes are made of lipid bilayers, where core can be hydrophobic or hydrophilic [53]. For instance, liposomes with aqueous core are used to trap aqueous soluble drug molecules, while core with hydrophobic core is utilized for lipid soluble drugs [17].

Liposomes are generally coated with polyethylene glycol [36, 53, 60]. For improvement of liposomes targeting capability and circulation time, it was coated with hyaluronan (HA) [10, 12]. Both passive and active targeting method can be used for liposome delivery. It can be conjugated with ligands or antibodies for selective drug delivery [15, 61]. Dendrimers are the branched 3-D tree-like structures with a multifunctional core; and these can be prepared by natural or synthetic elements, such as: sugars, nucleotides, and amino acids [58].

Preparation involves the controlled polymerization of the monomers maintaining desired size and shape. Hydrophilic and hydrophobic molecules can be conjugated to the branching of dendrimers [17, 22, 37]. Molecules are loaded in the cavity with varying interactions, such as: hydrogen binding, chemical linkages, or hydrophobic interaction [17]. Micelles are spherical structures, in which core is of hydrophobic tail and outer surface is of hydrophilic head. It is the effective nanocarrier for lipid soluble drug molecules [17, 37].

Nano-sphere is composed of matrix system, in which therapeutic agent is distributed by attachment, entrapment or encapsulation. Targeting can be improved by conjugating ligand or antibodies on nanoparticle surface [47, 53]. CNT is the hollow form of NPs. Atoms can be trapped inside the tube-like structure; while ligands or antibodies are conjugated to the surface [17, 53]. CNT's can be modified by liking proteins, peptides, therapeutic agents, and nucleic acid to enhance their water solubility [4, 30]. Various types of polymers for nanoparticle preparation are: polyglycolic acid, poly-alkyl cyanoacrylates, polylactic acid (PLA), poly-caprolactone; and their copolymers are exploited due to their biodegradation and biocompatibility [2, 5]. Table 8.1 demonstrates different types of nanoparticle polymers and their application as therapeutic agents.

8.6   APPLICATIONS OF NANOTECHNOLOGY IN DRUG DELIVERY

The nanoparticle, Abraxane, is used in breast cancer treatment and non-small cell lung cancers. University of Texas MD Anderson Cancer Center and Rice University has successfully employed the NPs in a mice model for the treatment of neck and head cancer. They replaced the toxic Cremophor EL with carbon NPs for intravenous injection of hydrophobic

TABLE 8.1 Polymeric Nanoparticles and Their Medical Outcomes

Polymeric Nanoparticle	Therapeutic Agent	Targeting Agent	Research Outcome	References
Polylactic acid	Paclitaxel	Monoclonal antibody (anti-HER2)	Enhanced breast cancer selective targeting	[7]
Poly-amidoamine	—	Folic acid	Increased cellular uptake	[59]
Polyethylene glycol	Small interfering RNA (siRNA)	Luteinizing hormone-releasing hormone (LHRH) peptide	Improved specificity	[56]
PEG-co-poly (lactic-co-glycolic acid)	Doxorubicin	Folic acid	Enhanced cytotoxicity and cellular uptake	[5]
PEG-poly (aspartate hydrazone doxorubicin)			Increased endocytotic cellular uptake	
Poly(D,L-lactidecoglycolide)	Cystatin	Cytokeratin specific monoclonal antibody	Prevent metastasis	[24]
Nanoshell (biodegradable polymer)	Docetaxel	Folic acid	Targeted and controlled delivery	[29]
Poly(D,L-lactidecoglycolide)	Paclitaxel	Folic acid	Improved inhibition of P-glycoprotein	[43]
Polyethylene glycol and polylactic acid			Increased drug accumulation in cancer cells	[44]

paclitaxel that resulted in improved drug targeting and reduced paclitaxel dose requirement [41].

Another example is use of nanoparticle chain in delivery of doxorubicin for breast cancer treatment in a mice model at Case Western Reserve University. Long nanoparticle chain was prepared using iron-oxide nano-spheres, three magnetic to one doxorubicin-loaded liposome. When magnetic nanoparticles (MNPs) entered into the cancer cells; then by vibrating them to generate radio-frequency resulted in the liposome rupture, hence drug was released inside the tumor cells. Growth of tumor was ceased by the nanotechnology than the standard treatment with doxorubicin and was reported to be safe as very low doxorubicin dose was required [46].

Scientists from Harvard University-Wyss Institute have demonstrated use of the biomimetic strategy in a mouse model. Drug coated NPs were used to disintegrate the blood clots by targeted binding to the narrowed regions in the blood vessels. Nanoparticle coated with tissue plasminogen activator dissolved the blood clots when injected intravenously. The nano-therapy has helped in reduction of bleeding that mainly occurs in thrombosis treatment [23].

Minicell NPs are employed in initial phase of clinical trial for drug delivery in patients for the treatment of advanced staged cancer. In general, minicell was prepared from the mutant bacteria membranes and were loaded with paclitaxel and coated with antibodies, cetuximab for the treatment of various types of cancer. This drug delivery system uses less drug doses and hence had low-side effects; and due to loading of multiple drugs, it has potential to treat different types of cancer [41]. Nano-sponges can bind to poorly soluble drugs and improve their bioavailability, due to their porous and small size. They can be used for targeting specific sites, prevent proteins and drugs from degradation, and help in controlled release of drug [1].

8.7 SUMMARY

This chapter briefly discusses nanotechnology applications in targeted cancer therapy. Nanotechnology came up as a promising technique for treatment and diagnosis of cancer at molecular level. In first section, different nanotechnology approaches towards cancer therapy were discussed. Later, active, and passive delivery systems were discussed with brief explanation.

Followed by biosynthesis of different delivery vehicles and application of nanotechnology, drug delivery equipment was presented.

KEYWORDS

- **bioavailability**
- **biosynthesis**
- **delivery vehicles**
- **liposome**
- **nanoparticles**
- **targeted delivery**

REFERENCES

1. Ahmed, R. Z., Patil, G., & Zaheer, Z., (2013). Nanosponges: Completely new nano-horizon: Pharmaceutical applications and recent advances. *Drug Development and Industrial Pharmacy, 39*(9), 1263–1272.
2. Barratt, G., (2003). Colloidal drug carriers: Achievements and perspectives. *Cellular and Molecular Life Sciences CMLS, 60*(1), 21–37.
3. Bharali, D. J., & Mousa, S. A., (2010). Emerging nanomedicines for early cancer detection and improved treatment: Current perspective and future promise. *Pharmacology & Therapeutics, 128*(2), 324–335.
4. Bianco, A., Kostarelos, K., & Prato, M., (2005). Applications of carbon nanotubes in drug delivery. *Current Opinion in Chemical Biology, 9*(6), 674–679.
5. Brewer, E., Coleman, J., & Lowman, A., (2011). Emerging technologies of polymeric nanoparticles in cancer drug delivery. *J. Nanomaterials, 2011*, 1–10.
6. Cho, K., Wang, X., Nie, S., Chen, Z., & Shin, D. M., (2008). therapeutic nanoparticles for drug delivery in cancer. *Clinical Cancer Research, 14*(5), 1310–1316.
7. Cirstoiu-Hapca, A., Buchegger, F., Bossy, L., Kosinski, M., Gurny, R., & Delie, F., (2009). Nanomedicines for active targeting: Physico-chemical characterization of paclitaxel-loaded Anti-HER2 immuno nanoparticles and *in vitro* functional studies on target cells. *European Journal of Pharmaceutical Sciences, 38*(3), 230–237.
8. Coates, A., Abraham, S., Kaye, S. B., Sowerbutts, T., Frewin, C., Fox, R. M., & Tattersall, M. H. N., (1983). On the receiving end; patient perception of the side-effects of cancer chemotherapy. *European Journal of Cancer, 19*(2), 203–208.
9. Dai, L., Liu, K., Si, C., Wang, L., Liu, J., He, J., & Lei, J., (2016). ginsenoside nanoparticle: New green drug delivery system. *Journal of Materials Chemistry B, 4*(3), 529–538.

10. Dan, P., & Rimona, M., (2004). Loading mitomycin C inside long circulating hyaluronan targeted nano-liposomes increases its antitumor activity in three mice tumor models. *International Journal of Cancer, 108*(5), 780–789.

11. Duncan, R., (2006). Polymer conjugates as anticancer nanomedicines. *Nature Reviews Cancer, 6,* 688–693.

12. Eliaz, R. E., & Szoka, F. C., (2001). Liposome-encapsulated doxorubicin targeted to CD44. A strategy to kill CD44-overexpressing tumor cells. *Cancer Research, 61*(6), 2592–2601.

13. Ferrari, M., (2005). Cancer nanotechnology: Opportunities and challenges. *Nature Reviews Cancer, 5,* 161–170.

14. Ganeshkumar, M., Ponrasu, T., Raja, M. D., Subamekala, M. K., & Suguna, L., (2014). Green synthesis of pullulan stabilized gold nanoparticles for cancer targeted drug delivery. *Spectrochimica Acta Part A: Molecular and Biomolecular Spectroscopy, 130,* 64–71.

15. Grobmyer, S. R., Zhou, G., Gutwein, L. G., Iwakuma, N., Sharma, P., & Hochwald, S. N., (2012). Nanoparticle delivery for metastatic breast cancer. *Nanomedicine: Nanotechnology, Biology and Medicine, 8,* S21–S30.

16. Guo, X., & Szoka, F. C., (2003). Chemical approaches to triggerable lipid vesicles for drug and gene delivery. *Accounts of Chemical Research, 36*(5), 335–341.

17. Haley, B., & Frenkel, E., (2008). Nanoparticles for drug delivery in cancer treatment. *Urologic Oncology: Seminars and Original Investigations, 26*(1), 57–64.

18. Jabir, N. R., Tabrez, S., Ashraf, G. M., Shakil, S., Damanhouri, G. A., & Kamal, M. A., (2012). Nanotechnology-based approaches in anticancer research. *International Journal of Nanomedicine, 7,* 4391–4408.

19. Jain, R. K., & Stylianopoulos, T., (2010). Delivering nanomedicine to solid tumor. *Nature Reviews Clinical Oncology, 7,* 653–660.

20. Jeengar, M. K., Rompicharla, S. V. K., Shrivastava, S., Chella, N., Shastri, N. R., Naidu, V. G. M., & Sistla, R., (2016). Emu oil based nano-emulgel for topical delivery of curcumin. *International Journal of Pharmaceutics, 506*(1), 222–236.

21. Jena, M., Mishra, S., Jena, S., & Mishra, S. S., (2017). *Nanotechnology- Future Prospect in Recent Medicine: A Review, 2*(4), 7–13.

22. Kojima, C., Kono, K., Maruyama, K., & Takagishi, T., (2000). Synthesis of poly-amidoamine dendrimers having poly(ethylene glycol) grafts and their ability to encapsulate anticancer drugs. *Bioconjugate Chemistry, 11*(6), 910–917.

23. Korin, N., Gounis, M. J., Wakhloo, A. K., & Ingber, D. E., (2015). targeted drug delivery to flow-obstructed blood vessels using mechanically activated nanotherapeutics. *JAMA Neurology, 72*(1), 119–122.

24. Kos, J., Obermajer, N., Doljak, B., Kocbek, P., & Kristl, J., (2009). Inactivation of harmful tumor-associated proteolysis by nanoparticulate system. *International Journal of Pharmaceutics, 381*(2), 106–112.

25. Krishna, R., & Mayer, L. D., (2000). Multidrug resistance (MDR) in cancer: Mechanisms, reversal using modulators of MDR and the role of MDR modulators in influencing the pharmacokinetics of anticancer drugs. *European Journal of Pharmaceutical Sciences, 11*(4), 265–283.

26. Kumar, C. G., & Poornachandra, Y., (2015). Biodirected synthesis of miconazole-conjugated bacterial silver nanoparticles and their application as antifungal agents and drug delivery vehicles. *Colloids and Surfaces B: Biointerfaces, 125,* 110–119.

27. Kumar, D. L., Mitra, J., & Roopa, S., (2020). Nanoencapsulation of food carotenoids. Chapter 7. In: Ragan, S., Nandita, D., & Eric, L., (eds.), *Environmental Nanotechnology* (Vol. 3, pp. 203–242). Cham: Springer.

28. LaVan, D. A., McGuire, T., & Langer, R., (2003). Small-scale systems for *in vivo* drug delivery. *Nature Biotechnology, 21*, 1184–1192.

29. Liu, Y., Li, K., Pan, J., Liu, B., & Feng, S. S., (2010). Folic acid conjugated nanoparticles of mixed lipid monolayer shell and biodegradable polymer core for targeted delivery of docetaxel. *Biomaterials, 31*(2), 330–338.

30. Liu, Z., Chen, K., Davis, C., Sherlock, S., Cao, Q., Chen, X., & Dai, H., (2008). Drug delivery with carbon nanotubes for *in vivo* cancer treatment. *Cancer Research, 68*(16), 6652–6660.

31. Lohith, K. D. H., & Sarkar, P., (2018). Encapsulation of bioactive compounds using nanoemulsions. *Environmental Chemistry Letters, 16*(1), 59–70.

32. Lu, R., Yang, D., Cui, D., Wang, Z., & Guo, L., (2012). Egg white-mediated green synthesis of silver nanoparticles with excellent biocompatibility and enhanced radiation effects on cancer Cells. *International Journal of Nanomedicine, 7*, 2101–2107.

33. Maeda, H., Wu, J., Sawa, T., Matsumura, Y., & Hori, K., (2000). Tumor vascular permeability and the epr effect in macromolecular therapeutics: A review. *Journal of Controlled Release, 65*(1), 271–284.

34. Malam, Y., Loizidou, M., & Seifalian, A. M., (2009). Liposomes and nanoparticles: Nanosized vehicles for drug delivery in cancer. *Trends in Pharmacological Sciences, 30*(11), 592–599.

35. Malathi, S., Balakumaran, M., Kalaichelvan, P., & Balasubramanian, S., (2013). Green synthesis of gold nanoparticles for controlled delivery. *Advanced Materials Letters, 4*(12), 933–940.

36. Matsumura, Y., Hamaguchi, T., Ura, T., Muro, K., Yamada, Y., Shimada, Y., Shirao, K., et al., (2004). Phase I clinical trial and pharmacokinetic evaluation of NK911, a micelle-encapsulated doxorubicin. *British Journal of Cancer, 91*, 1775–1785.

37. Moghimi, S. M., Hunter, A. C., & Murray, J. C., (2005). Nanomedicine: Current status and future prospects. *The FASEB Journal, 19*(3), 311–330.

38. Mousa, S. A., & Bharali, D. J., (2011). Nanotechnology-based detection and targeted therapy in cancer: Nano-bio paradigms and applications. *Cancers, 3*(3), 2888–2899.

39. Nagahara, L. A., Lee, J. S. H., Molnar, L. K., Panaro, N. J., Farrell, D., Ptak, K., Alper, J., & Grodzinski, P., (2010). Strategic workshops on cancer nanotechnology. *Cancer Research, 70*(11), 4265–4268.

40. Nie, S., Xing, Y., Kim, G. J., & Simons, J. W., (2007). Nanotechnology applications in cancer. *Annual Review of Biomedical Engineering, 9*(1), 257–288.

41. Nikalje, A. P., (2015). Nanotechnology and its applications in medicine. *Med. Chem., 5*(2), 81–89.

42. Park, K., (2007). Nanotechnology: What can it do for drug delivery. *Journal of Controlled Release, 120*(1), 1–3.

43. Patil, Y., Sadhukha, T., Ma, L., & Panyam, J., (2009). Nanoparticle-mediated simultaneous and targeted delivery of paclitaxel and tariquidar overcomes tumor drug resistance. *Journal of Controlled Release, 136*(1), 21–29.

44. Patil, Y. B., Toti, U. S., Khdair, A., Ma, L., & Panyam, J., (2009). Single-step surface functionalization of polymeric nanoparticles for targeted drug delivery. *Biomaterials, 30*(5), 859–866.

45. Peer, D., Karp, J. M., Hong, S., Farokhzad, O. C., Margalit, R., & Langer, R., (2007). Nanocarriers as an emerging platform for cancer therapy. *Nature Nanotechnology, 2,* 751–760.

46. Peiris, P. M., Bauer, L., Toy, R., Tran, E., Pansky, J., Doolittle, E., Schmidt, E., et al., (2012). Enhanced delivery of chemotherapy to tumors using a multicomponent nanochain with radio-frequency-tunable drug release. *ACS Nano, 6*(5), 4157–4168.

47. Pérez-Herrero, E., & Fernández-Medarde, A., (2015). Advanced targeted therapies in cancer: Drug nanocarriers, the future of chemotherapy. *European Journal of Pharmaceutics and Biopharmaceutics, 93,* 52–79.

48. Praetorius, N. P., & Mandal, T. K., (2007). Engineered nanoparticles in cancer therapy. *Recent Patents on Drug Delivery & Formulation, 1*(1), 37–51.

49. Pulicharla, R., Marques, C., Das, R. K., Rouissi, T., & Brar, S. K., (2016). Encapsulation and release studies of strawberry polyphenols in biodegradable chitosan nanoformulation. *International Journal of Biological Macromolecules, 88,* 171–178.

50. Singh, M., Kumar, M., Manikandan, S., Chandrasekaran, N., Mukherjee, A., & Kumaraguru, A., (2014). Drug delivery system for controlled cancer therapy using physico-chemically stabilized bioconjugated gold nanoparticles synthesized from marine macroalgae, *Padina gymnospora. Journal of Nanomedicine & Nanotechnology,* (S5), 1–11.

51. Storm, G., Belliot, S. O., Daemen, T., & Lasic, D. D., (1995). Surface modification of nanoparticles to oppose uptake by the mononuclear phagocyte system. *Advanced Drug Delivery Reviews, 17*(1), 31–48.

52. Sutradhar, K. B., & Amin, M. L., (2013). Nanoemulsions: Increasing possibilities in drug delivery. *European Journal of Nanomedicine, 5*(2), 97–110.

53. Sutradhar, K. B., & Amin, M. L., (2014). Nanotechnology in cancer drug delivery and selective targeting. *ISRN Nanotechnology, 2014,* 12–20.

54. Tallury, P., Kar, S., Bamrungsap, S., Huang, Y. F., Tan, W., & Santra, S., (2009). Ultra-small water-dispersible fluorescent chitosan nanoparticles: Synthesis, characterization and specific targeting. *Chemical Communications,* (17), 2347–2349.

55. Tannock, I. F., Lee, C. M., Tunggal, J. K., Cowan, D. S. M., & Egorin, M. J., (2002). Limited penetration of anticancer drugs through tumor tissue. *A Potential Cause of Resistance of Solid Tumors to Chemotherapy, 8*(3), 878–884.

56. Taratula, O., Garbuzenko, O. B., Kirkpatrick, P., Pandya, I., Savla, R., Pozharov, V. P., He, H., & Minko, T., (2009). Surface-engineered targeted PPI dendrimer for efficient intracellular and intratumoral siRNA delivery. *Journal of Controlled Release, 140*(3), 284–293.

57. Torchilin, V. P., & Trubetskoy, V. S., (1995). Which polymers can make nanoparticulate drug carriers long-circulating? *Advanced Drug Delivery Reviews, 16*(2), 141–155.

58. Wang, A. Z., Langer, R., & Farokhzad, O. C., (2012). Nanoparticle delivery of cancer drugs. *Annual Review of Medicine, 63*(1), 185–198.

59. Yang, W., Cheng, Y., Xu, T., Wang, X., & Wen, L. P., (2009). Targeting cancer cells with biotin-dendrimer conjugates. *European Journal of Medicinal Chemistry, 44*(2), 862–868.

60. Yezhelyev, M. V., Gao, X., Xing, Y., Al-Hajj, A., Nie, S., & O'Regan, R. M., (2006). Emerging use of nanoparticles in diagnosis and treatment of breast cancer. *The Lancet Oncology, 7*(8), 657–667.
61. Yiyao, L., Hirokazu, M., & Michihiro, N., (2007). Nanomedicine for drug delivery and imaging: A promising avenue for cancer therapy and diagnosis using targeted functional nanoparticles. *International Journal of Cancer, 120*(12), 2527–2537.
62. Zhao, G., & Rodriguez, B. L., (2013). Molecular targeting of liposomal nanoparticles to tumor microenvironment. *International Journal of Nanomedicine, 8*, 61–71.

PART III
Role of Nanotechnology in Bioprocessing

CHAPTER 9

NANOTECHNOLOGY FOR BIOFUEL PRODUCTION

NIKHIL KUMAR and PAWAN KUMAR

ABSTRACT

Current application of Nanotechnology in biofuel production is limited. This is due to number of limiting factors including technical and economic issues, which are present in the nanotechnologies and in the bioenergy industry. The potential application and scope of nanotechnology on biofuel production has been discussed in this chapter.

9.1 INTRODUCTION

In recent years, utilization of biofuel as an alternative to non-renewable fossil fuel has gained greater interest. Biofuels has shown their potential as eco-friendly and enhanced functional properties in comparison with petroleum-derived fuels. In addition, biodegradability, carbon neutrality, inherence, and high flash point are main functional characteristics of biofuels/biofuel-feed stocks. Nanotechnology is playing a major role in biofuel production and enhancement. Numerous nanoparticles (NPs) have been investigated for their application in biofuel production. For instance, nanoparticle-assisted enzyme hydrolysis of lignocellulosic biomass for the production of bioethanol has been investigated. Different metal NPs (such as: titanium dioxide and zinc oxide (ZnO)) are common nanostructures to immobilize industrially important enzymes.

Magnetic nanoparticles (MNPs) can immobilize the enzymes for lignocellulosic hydrolysis using different strategies, such as: covalent bonding, physical adsorption, and specific ligand receptors and cross-linking. These enzymes are generally referred as nano-catalysts, which are fabricated using functional NPs. In addition, in bioenergy and biofuel sector, nanotechnology has different applications in the development of efficient nano-catalysts *via* enhanced surface area, nano-membranes, and modification of feed stocks through pre-treatment.

For example, enzymes have been largely used to hydrolyze lignocellulosic biomass to produce ethanol and biogas or to catalyze biodiesel production from oils and fats [10, 16, 18]. In this context, nanomaterials can be used to replace the enzymes or to immobilize them, resulting into more efficient catalysis or favoring the recovery of biocatalysts from medium. Moreover, this technology includes alternatives, in which magnetic properties are added to immobilized systems. Most significant property of NPs is that they can be easily recovered from reaction mixture by applying suitable magnetic field [1]. This chapter focuses on role, scope, and applications of Nanotechnology on biofuel production.

9.2   NANOTECHNOLOGY USE IN BIOFUEL PRODUCTION

9.2.1   NANOTECHNOLOGY IN FEED-STOCK PRE-TREATMENT

One of the nanotechnology aspects in biofuel production is development and processing of new materials/catalysis at nanoscale, which is one of the most recent subjects that has garnered great interest from the scientific community.

A significant aspect in biofuel production is the nanoscale or sub-nanoscale instrumentation, which contributed to the understanding of the cell-wall ultrastructure and the mechanisms of enzymatic hydrolysis. In the long-term, this aspect of nanotechnology contributes to biofuel production, advances in instrumentation, and methods for sample preparation. Nanotechnology can also be one of the solutions for lignocellulosic residue accumulation, which could be used in biofuels and biomaterial production. Proper use of residue could highly reduce its accumulation, and the transformed products could generate extra income.

Lignocellulosic biomass are potential feedstocks for production of biofuel due to their abundant carbon source. However, lignocellulosic biomass has been utilized at industrial scale to generate the energy by burning as self-sustainable energy process. Nevertheless, lignocellulosic biomass comprises of cellulose and hemicellulose as polymeric carbohydrates, which can be valorized to value-added products, such as: bio-oils, bio-alcohol, and other industrially essential chemicals through bioprocessing. In this context, nanotechnology demonstrates significant potential to improve the biofuel production by using different types of nanostructures. Nanomaterials possess unique physical, electrical, and mechano-chemical properties, which differ from their bulk material. The multitude of nanotechnology helps in understanding at molecular level and bioconversion of lignocellulosic biomass.

Acid functionalized NPs and nanoscale shear hybrid techniques are considered as a part of a feed stock pre-treatment using nanotechnology principles. In acid functionalized NPs technique, different solid acid nano-catalysts with ability to catalyze hydrolysis reactions are considered. It has been demonstrated in recent studies that MNPs functionalized by acid treatment possess similar catalytic effect on mineral acids, which are used in different chemical pre-treatment methods. In addition, due to their recoverable and reusable magnetic property, they reduce the downstream processing cost and hence they are potential candidates for lignocellulose biomass (LCB) pretreatment. In a recent study, Lai et al. demonstrated the potential of sulfonated silica MNPs composite to hydrolyze the lignin biomass. These nanocomposites are functionalized with iron oxide to induce the magnetic property.

Methods for treatment of LCB are physical, chemical, biological, biochemical, and other pretreatment method. Physical treatment method consists of mechanical extrusion using milling and grinding equipment and then scarification of cellulose and hemicellulose into simple sugar (glucose, xylose, arabinose, etc.), using microwave or ultrasonic; because it creates small bubble cavitation and thus increasing the accessible surface area for enzymatic treatment. While in chemical treatment, high concentration of acid and alkaline is used for scarification of cellulose and hemicellulose, and provide glucose and xylose without losing the lignin compound in the LCB. However, when nanotechnology used in the pretreatment of LCB, it becomes eco-friendly and economic viable method.

Nano biocatalyst is more helpful in contrast of economic and purification. For the treatment of LCB when acid and immobilized enzyme is functionalized through magnetic nanomaterial, then it is a breakthrough alternative to conventional methods. By using this, same enzyme can be recyclable many times. There are some advantages using NPs in pretreatment [5], such as:

- Continuous process gives more retention time to LBS for reacting with enzymes and get a very high conversion;
- It is a fast process, which completes in few minutes; so overall cost of the process is low;
- Need of simple chemicals, such as: for pretreatment sulfuric Acid and for neutralizing sodium hydroxide, there is very less use of harsh chemicals, which affect the environment [5];
- There is no requirement of high heat and pressure, so it needs low energy.

9.2.2   NANOTECHNOLOGY IN DETOXIFICATION OF BIOMASS

Using nanotechnology in downstream processing can help in improving the energy efficiency and it can detect and remove the environmental contaminates. It can provide affordable water treatment using nano-technology. Molybdenum disulfide membrane filters are more efficient than current conventional filter, because it has thin film membrane with Nano pores for optimum energy consumption. Nanoparticle can also treat industrial waste water through chemical reaction in underground so there is no need to pump water outside. Acid, which is used during pretreatment of LCB, is harmful and it releases openly into the environment and also it increases the cost of production of biofuel. Therefore, scientists are thinking to reduce the cost by acid recycling, which removes the acid from Lignocellulose Hydrolysate for fermentation; and that removed acid is also collected for further pretreatment of biomass. Acid is removed using ion exchange membrane through different techniques:

1.   **Electrodialysis:** There is constant current flowing through different chambers and electro dialysis unit and separate the acid from concentrate compartment to dilute compartment.
2.   **Diffusion Dialysis:** Acid is separated through concentration gradient only.

9.2.3   NANOTECHNOLOGY IN FERMENTATION

Due to limited quality and high economic value of nanomaterials, it creates problem in exploitation for end-user and investors also do not want to take risks to invest in this new technology because its demand and market value is not much clearly defined. First time nanoscale inorganic engineering material was used by Tommy Phelps, instead of organic compounds using bioprocess technique, in Oak Ridge National Laboratory.

Nano fermentation hampered the natural metabolic pathways and induced the new pathway, which can be used to make specific engineering material in metal reducing bacteria. For example, in metal reducing bacteria having the metabolic process suited for anaerobic conditions (which convert Fe(III) to Fe(II) in the presence of electron donor (Hydrogen or Glucose)), electron transfer reaction happens because of metal reductases enzymes. It was also assumed in past that bacteria can only produce pure metal. Some metal reducing bacteria can produce metal outside the membrane, so that it can be easily harvested without damaging the cell; and next time the same bacteria with different feedstock can produce mixed metal.

Nanofermentation process can be used for bacteria culture at low temperature (*Shewanella alga*) and high temperature (*Thermoanaerobacter ethanolicus*). NPs have been produced through fermentation using metal reducing bacteria and complex metal oxides (Zn, Co, Ni, Mn, Pd) and rare earth metal oxides (neodymium, gadolinium, erbium, and terbium), uranium. Other applications of nanofermentation are doped iron metal in nanoscale particle, because each single particle having the properties of ferromagnetic domain, paramagnetic, and Ferro fluids [4]. Researchers have demonstrated continuous fermentation process for the production of bioethanol, where immobilized *S. cerevisiae* cells showed high ethanol production capability [6].

Nanotechnology is also employed in different types of sensors, which is used for detection of biological and chemical compounds; and it also helps in controlling the parameters of fermentation for better productivity. Biosensor converts biological signal to electrical signal through the transducer, which can detect the signal due to electric potential.

The examples of different types of biosensors are: electronic tongue, microbial sensor, and electrochemical apta-sensor. The nanosensors can be classified as [3]:

- Carbon nanotubes (CNTs);
- Colloidal gold (Au) nanoparticles;
- Magnetic nanoparticles (MNPs);
- Silver (Ag) nanoparticles;
- Zinc oxide nanoparticles (ZnO NPs).

Kadar et al. supplemented three microalgae (*Pavlova lutheri, Isochrysis galbana, and Tetraselmis suecica*) with synthetic nanoscale zero-valent iron NPs. They observed that *Pavlova lutheri and Tetraselmis suecica* accumulated higher lipid content in comparison to the cultures supplemented with standard Fe-EDTA. Higher lipid accumulation in algae was in response to the oxidative stress caused by zero-valent iron NPs generated reactive oxygen species (ROS) [7].

Another research group reported an increased cell growth of green microalga *Chlorella vulgaris* incubated with 38–190 nm sized laser-ablation-prepared silica NPs [12]. Research study on *Chlorella vulgaris* UTEX 265 indicated a slight increase in lipid content with the supplementation of TiO_2 NPs (@ 0.1 g/L) under UV treatment. The growth of alga, however, was decreased at 2.5 and 5.0 g/L of TiO_2 NPs, demonstrating the negative effect of higher concentration of TiO_2 NPs in the presence of UV radiation [8].

9.2.4 NANOTECHNOLOGY IN DOWNSTREAM PROCESSING

9.2.4.1 SEPARATION AND RECOVERY OF PRODUCTS

Nanofiltration process is used to separate divalent and multivalent ions and permeates the particle, which is <0.001 microns in size. It is far better than reverse osmosis (RO) in economic comparison. Nanofiltration is used in demineralization, milk, whey, desalination, and juice filtration. In current scenario for separation of proteins, nucleic acids and viral vaccines, adsorptive, and chromatographic techniques are used.

Adsorptive process contains different types of high-performance liquid chromatography (HPLC), which cannot handle cell debris and other colloidal particles. This problem can be solved by filtration process; but filtration takes more time because of low flux rate and membrane fouling. In adsorption process it can be slow, because when large salutes are targeted through beads using matrix it significantly decreases the active surface area. It is concluded that only outer side is effective for protein capture, therefore it can suffer capacity limitations.

In this separation technique, there is use of magnetic fluid NPs, which are recovered using High gradient magnetic separation technology. When magnetic field is applied to these particles then there is a direction behavior due to their polarity and also it puts some force on particles. One limitation of this separation is that some time magnetic field is insufficient to take the particles and hold them against the diffusive and drag forces in the column. To solve this problem, the channel is packed with magnetic steel wool that de-homogenizes the magnetic field by distorting the field lines near the surface; and provides a high magnetic field gradient to allow filtration from the solution.

9.2.4.2   NANOFILTRATION

Nanofiltration is a pressure driven process applied in the separation of ions from solute, such as, small molecules of sugars. This type of membrane opens new opportunities for improving the process efficiency and production of new products. Pressure applied to it is generally less than RO pressure. When pretreatment is done at high temperature, there is high chance of production of furfural (which is an inhibitory compound for fermentation) that can be removed using Nanofiltration membrane and can also concentrate the sugars, which are found in Hydrolysate [11]. During fermentation, sometimes acetic concentration is so high that it can inhibit the growth of microorganisms and can also inhibit product formation; acetic acid can be removed by using Nanofiltration membrane and the productivity of fermentation product can be improved [20].

Nanofiltration can be used with high pressure in bioethanol separation [9]. One of the early studies regarding the usage of nanofiltration for ethanol and sugars separations was carried out by Verhoef research group, who used hydrophobic nanofiltration membrane for the separation of ethanol from multicomponent mixtures [17]. Bras and his research group used three nanofiltration membranes NF270, NF90, and SW30, of which NF270 was most efficient in the separation of bioethanol from the fermented liquors of olive stones. This nanofiltration membrane showed 98% sugar rejection and 28% of lower ethanol rejection, which indicated that this separation membrane was ideal for recovery of bioethanol from such sugars [4].

Recently Shibuya and his group used hybrid of nano-filtration and forward osmosis technique successfully for enhancing bioethanol concentration from xylose-assimilating S. cerevisiae, whose liquid fraction after dilution was 1.5-folds. This hybrid system was useful in the removal of renowned fermentation inhibitors, such as: acetic acid. Such hybrid nanofiltrations systems are of significant potency for efficient separation of bioethanol from pretreated lignocellulosic biomass [13].

9.2.5   NANO-ADDITIVES IN FUEL BLENDING

Biodiesel is an alternate option for extending the value of fossil fuels and also enduring and asepsis of engine. It reduces the carbon emission and vulnerability on fossil fuels. Examples of issues related with this fuel are: (1) incompatibility with cold weather; (2) increase in the NO_2 emission; and (3) increase in the maintenance cost of engine (such as: fuel filters, fuel tanks and fuel lines) due to clogging. Fuel properties can be increased by mixing with NPs as fuel additives (e.g., metallic, non-metallic, oxygenated, organic, and combination), which are stable thermally, better heat transfer rate and result in stabilization of fuel mixture. Amount of Nano additives will increase the engine performance and reduce carbon emission. When Nano-additives are blended with biodiesel, it increases the engine efficiency without changing the engine technical properties.

Nano-additives are prepared through Al and Ag NPs along with respective surfactants. Its high production cost is main barrier to the application of NPs in the biofuel industry; and production method requires very high and advanced equipment. Conclusion can be drawn from different type of nano-additives with respective surfactant:

- Alteration to biodiesel can be made to improve efficiency and off gasses characteristics of engine;
- Blended biodiesel can accelerate the evaporation, which decreases the ID and CD, while higher PP and HRR are observed;
- Brake thermal efficiency (BTE) is improved with biodiesel blended with nanoparticles while exhaust gasses (CO, HC, and NOx) are reduced as compared to biodiesel alone.

When these NPs are blended with biodiesel, this mixture will increase the combustion characteristics. The main problem with addition of Nano additives is their stability. The nanoparticle will make clump because of large surface area. For preventing this issue, there is need of a surfactant to reduce the coagulation and agitation of nanoparticle in the liquid fuel.

9.2.5.1   STEPS INVOLVED IN THE STUDY OF NANOPARTICLE-DIESEL BIODIESEL EMULSION FUEL

- Collect biodiesel and nanoparticles in different sources, such as, biodiesel from Dairy scum and NPs from Sigma Aldrich Company;

- Perform two-unit operation blending and heating of NPs and Biodiesel;
- Use ultrasonicator for mixing at fixed temperature;
- Prepare (biodiesel and various concentration of nanoparticles) a solution in different bottles;
- Check the stability of nanoparticles in biodiesel through the value of zeta potential;
- For checking the fuel characteristics, calculate Octane number, Calorific value, etc.;
- Check the engine performance in fuel using BTE, BP (brake power), BSFC;
- Check the composition of exhaust gasses: CO, CO_2, NO_x, PM, and SO_x.

9.2.5.2 PERFORMANCE ANALYSIS OF NANO-ADDITIVES WITH DIESEL-BIODIESEL FUEL BLEND

Biodiesel has an alternate option of fossil fuel as a green energy. Having approximately same properties as fossil fuel and few drawbacks, it is less useful as a fuel. These drawbacks can be reduced by adding nano-additives, which can affect following performance parameters of the engine, such as:

- Brake thermal efficiency (BTE) is the energy difference between energy of the fuel and energy generated by engine;
- Brake specific fuel consumption (BSFC) is the ratio of volume expansion of fuel with respect to time by the engine to the power generated with respect to time by the engine;
- Brake horsepower (BP) is the power output from shaft of an engine lacking any power loss due to gear and others friction.

9.2.5.3 ENGINE EMISSION CHARACTERISTICS OF DIESEL-BIODIESEL FUEL BLENDS WITH NANO-ADDITIVES

In diesel engines, there is less emission of hydrocarbon (HC) and carbon mono oxide (CO) but PM level is so high due to nitrogen oxide in comparison to petrol engine:

- Amount of biodiesel is indirectly proportional to the emission of CO; therefore, addition of Nano additives in biodiesel makes it more compressible in the presence of oxygen.

- When metal oxides added as Nano additives at different speeds, concentrations, and conditions, then there is a reduction in the emission of CO_2 from the engine.
- Hydrocarbons (HC) are the fragments of burned molecules of fuel. When nanoparticles are added in fuel, then it creates space for oxygen and optimize the ratio of air-fuel mixture thus helping in complete combustion of the fuel.
- Catalytic effect of Nano additives is to increase its combustion quality and to minimize the smoke formation, which reduces the particulate matter (PM) emission.
- By adding oxygenated additives in biofuel, there is reduction in the emission of oxides of nitrogen, such as [14]: NO (nitric oxide), NO_2 (nitrogen dioxide), which found in the exhaust pipe.

9.2.6 NANO-CATALYSTS

Several functionalized nanomaterials have been successfully used in the production of biodiesel and bioethanol. For example, sulfamic acid-functionalized nanocatalysts showed comparatively more activity and better performance [19]. In another investigation, researchers investigated the production of biodiesel from cooking oils, using CaO and MgO NPs synthesized by sol-gel and sol-gel self-combustion methods, respectively. CaO NPs showed significant increase in the biodiesel yield compared to MgO NPs.

Other interesting approach is the use of magnetic nanocatalysts that can be easily recovered and reused, favoring the economic viability of the process [15]. In this context, another research group presented significant potential toward hydrolysis, transesterification, and esterification of soybean oil and their fatty acids [2].

9.3 SUMMARY

Pretreatment is the important process for getting the valuable products from Lignocellulosic biomass but using NPs treatment will be more efficient and also eco-friendly. NPs can be used for blending with biodiesel to increase the cetane number of biodiesel and improve the efficiency of biodiesel and also to reduce the exhaust harmful gasses and to support clean and green energy. For treatment of LBS acid, functionalized Nanomaterials are used, which are recyclable after treatment. Addition of Nano-additives to biofuel helps to: reduce the CO_2 and other harmful gasses emission; increase the oxygen density in engine and filter; improve the fuel stability over wide operating conditions; optimum viscosity; and reduction in ignition delay time flash point.

KEYWORDS

- biofuels
- biomolecules
- brake thermal efficiency
- fermentation
- hydrocarbon
- nano-additives

REFERENCES

1. Ahmed, M., & Douek, M., (2013). The role of magnetic nanoparticles in the localization and treatment of breast cancer. *BioMed Research International, 2013,* 9. Article ID: 281230.
2. Alves, M. B., Medeiros, F. C. M., Sousa, M. H., Rubim, J. C., & Suarez, P. A. Z., (2014). Cadmium and tin magnetic nanocatalysts useful for biodiesel production. *Journal of Brazilian Chemical Society, 25,* 2304–2313.
3. Antonio, M., Iris, L., & Antonio, S. L. J., (2016). Nanotechnology and wine. In: *Novel Approaches of Nanotechnology in Food* (pp. 165–199). London: Elsevier.
4. Brás, T., Fernandes, M. C., Santos, J. L. C., & Neves, L. A., (2013). Recovering bioethanol from olive bagasse fermentation by nanofiltration. *Desalination and Water Treatment, 51*(22–24), 4333–4342.
5. Ingle, A. P., Chandel, A. K., Antunes, F. A., Rai, M., & Da Silva, S. S. (2019). New trends in application of nanotechnology for the pretreatment of lignocellulosic biomass. *Biofuels, Bioproducts and Biorefining, 13*(3), 776–788.
6. Ivanova, V., Petrova, P., & Hristov, J., (2011). Application in the ethanol fermentation of immobilized yeast cells in matrix of alginate/magnetic nanoparticles, on chitosan-magnetite microparticles and cellulose-coated magnetic nanoparticles. *Int. Rev. Chem. Eng., 3*(2), 289–299.

7. Kadar, E., Rooks, P., Lakey, C., & White, D. A., (2012). The effect of engineered iron nanoparticles on growth and metabolic status of marine microalgae cultures. *Science of the Total Environment, 439,* 8–17.

8. Kang, N. K., Lee, B., Choi, G. G., Moon, M., Park, M. S., Lim, J., & Yang, J. W., (2014). Enhancing lipid productivity of *Chlorella vulgaris* using oxidative stress by TiO_2 nanoparticles. *Korean Journal of Chemical Engineering, 31*(5), 861–867.

9. Kang, Q., Appels, L., Tan, T., & Dewil, R., (2014). Bioethanol from lignocellulosic biomass: Current findings determine research priorities. *The Scientific World Journal, 2014,* 11. Article ID: 298153.

10. Michalska, K., Bizukojć, M., & Ledakowicz, S., (2015). Pretreatment of energy crops with sodium hydroxide and cellulolytic enzymes to increase biogas production. *Biomass and Bioenergy, 80,* 213–221.

11. Qi, B., Luo, J., & Chen, X., (2011). Separation of furfural from monosaccharides by nanofiltration. *Bioresource Technology, 102*(14), 7111–7118.

12. San, N. O., Kurşungöz, C., Tümtaş, Y., Yaşa, Ö., Ortaç, B., & Tekinay, T., (2014). Novel one-step synthesis of silica nanoparticles from sugarbeet bagasse by laser ablation and their effects on the growth of freshwater algae culture. *Particuology, 17,* 29–35.

13. Shibuya, M., Sasaki, K., Tanaka, Y., Yasukawa, M., Takahashi, T., Kondo, A., & Matsuyama, H., (2017). Development of combined nanofiltration and forward osmosis process for production of ethanol from pretreated rice straw. *Bioresource Technology, 235,* 405–410.

14. Soudagar, M. E. M., Nik-Ghazali, N. N., Kalam, M. A., Badruddin, I., Banapurmath, N., & Akram, N., (2018). The effect of nano-additives in diesel-biodiesel fuel blends: A comprehensive review on stability, engine performance and emission characteristics. *Energy Conversion and Management, 178,* 146–177.

15. Tahvildari, K., Anaraki, Y. N., Fazaeli, R., Mirpanji, S., & Delrish, E., (2015). The study of CaO and MgO heterogenic nano-catalyst coupling on transesterification reaction efficacy in the production of biodiesel from recycled cooking oil. *Journal of Environmental Health Science and Engineering, 13*(1), 73–80.

16. Terán-Hilares, R., Reséndiz, A. L., Martínez, R. T., Silva, S. S., & Santos, J. C., (2016). Successive pretreatment and enzymatic saccharification of sugarcane bagasse in a packed bed flow-through column reactor aiming to support biorefineries. *Bioresource Technology, 203,* 42–49.

17. Verhoef, A., Figoli, A., Leen, B., & Bettens, B., (2008). Performance of a nanofiltration membrane for removal of ethanol from aqueous solutions by pervaporation. *Separation and Purification Technology, 60*(1), 54–63.

18. Verma, M. L., & Barrow, C. J., (2015). Recent advances in feedstocks and enzyme-immobilized technology for effective transesterification of lipids into biodiesel. In: Kalia, V. C., (ed.), *Microbial Factories: Biofuels, Waste treatment* (Vol. 1, pp. 87–103). V.C. New Delhi: Springer India.

19. Wang, H., Covarrubias, J., Prock, H., Wu, X., Wang, D., & Bossmann, S. H., (2015). Acid-functionalized magnetic nanoparticle as heterogeneous catalysts for biodiesel synthesis. *The Journal of Physical Chemistry C, 119*(46), 26020–26028.

20. Weng, Y. H., Wei, H. J., Tsai, T. Y., Chen, W. H., Wei, T. Y., Hwang, W. S., Wang, C. P., & Huang, C. P., (2009). Separation of acetic acid from xylose by nanofiltration. *Separation and Purification Technology, 67*(1), 95–102.

CHAPTER 10

FABRICATION AND APPLICATIONS OF ENZYME NANOPARTICLES

PAWAN KUMAR and NIKHIL KUMAR

ABSTRACT

Enzymes are widely utilized in food and pharmaceutical industries due to their catalytic activity that can be monitored and controlled by specific substrate, pH, temperature, concentration, and inhibitors. The increased surface area, particle mobility, thermal, and storage stability, diffusion, and easy dispersion are advantages of enzyme nanoparticles (ENPs) over the conventional enzymes. In addition, nanoparticles (NPs) can interact with bio-macromolecules due to resemblance in their size, which helps in fabrication of biosensors, enhancement of solubility and interaction with enzymes. This chapter summarizes recent developments in ENPs fabrication and their potential application in different bioprocessing field.

10.1 INTRODUCTION

Enzymes are globular biomolecules to catalyze biochemical reactions. In general, enzymes are synthesized inside the cells and secreted extracellularly or retained inside the cells depending on their function. Enzymes for industrial applications are produced by microorganisms, such as: bacteria, fungi, and archaea. The enzymes have wide applications in food industry, pulp, and paper industry, pharmaceutical, and cosmetics to produce value-added products (Figure 10.1).

The size of the enzymes varies between 10 nm to micrometers compared with 10–100 nm for enzyme nanoparticles (ENPs). The advancement in material functionalization at nanoscale has become the recent research trend [21]. At nanoscale, the optical, physical, mechanical, magnetic, electrical, chemical, and all the properties of materials can be altered according to the end-use. This surface functionalization has given advantage to nanomaterials for their industrial applications. There are thousands natural enzymes but only few of them have been prepared as NPs and characterized to utilize at the industrial scale. Therefore, production of ENPs is an emerging area of interest in research.

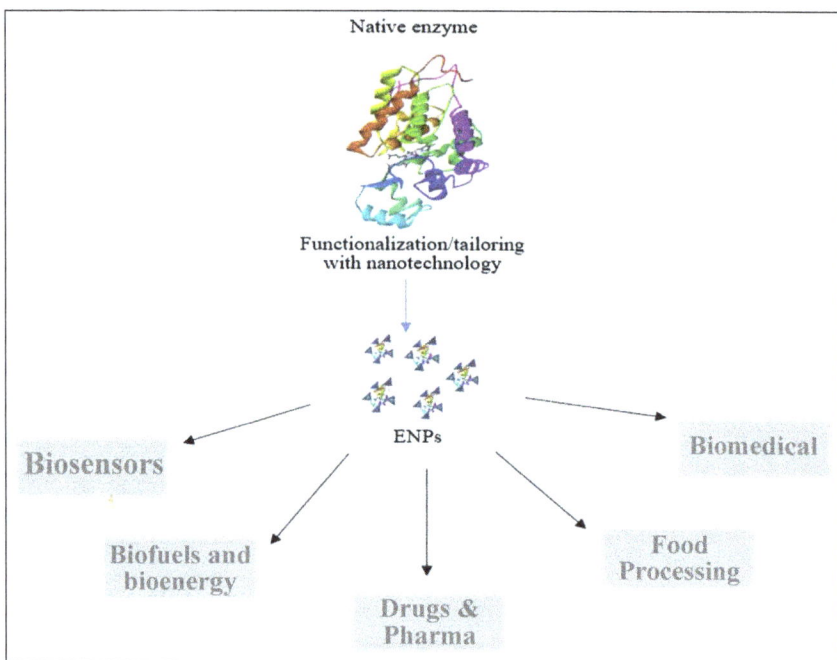

FIGURE 10.1   Potential applications of enzyme nanoparticles.

ENPs are clustered assembly of single or multiple enzyme molecule in the form of protein matrix at nanoscale. Tailoring of enzymes at nanoscale provides different advantages (such as: enhanced electrical, optical, chemical, thermal, magnetic, and catalytic activity) due to increased ratio of surface area to volume. In addition, nano-tailoring of enzymes offers enhanced performance of enzyme-based biosensors [14]. However, direct interaction between nano-template and enzymes can lead to decreased activity of an enzyme due to denaturation. Hence, to prevent denaturation of enzyme molecules, they are aggregated at nanoscale and self-assembled at controlled rate before immobilization.

This technique of immobilization at nanoscale has proven the potential for development of novel biosensors with enhanced detection limit and analytical performance. For instance, different ENPs/ bio-sensors (such as: horse radish peroxidase [14], glucose oxidase (GOx) [12, 21] and uricase [3]) have been developed. The ENPs are different from single ENPs, wherein single ENPs are functionalized/ armored at nanoscale with enhanced thermal stability and catalytical activity (such as: trypsin and chymotrypsin NPs with improved enzyme activity and thermal stability [11]). In a recent research study, horse radish peroxidase and cellulase also demonstrated the enhanced enzyme stability at elevated temperature and degradation efficacy.

This chapter focuses on the recent developments in preparation and application of nano-enzyme molecules/ENPs in bioprocessing, such as: food, pharma, pulp, and paper, textile, and bioprocessing.

10.2   PREPARATION AND CHARACTERIZATION OF ENZYME NANOPARTICLES (ENPS)

ENPs are prepared by biochemical interaction between NPs and protein molecules or protein macro-molecules itself. Emulsification, de-solvation, and coacervation are widely used strategies in enzyme nanoparticle preparation. For example, enzyme nanoparticle can be formed by aggregating the enzyme with ethanol droplets at 4°C that causes dehydration around the protein molecule to increase the molecular forces (such as: electrostatic and hydrophobic interaction and *Van der Waals* forces), which result in formation of ENPs.

In the emulsification method, emulsion of albumin aqueous solution at room temperature is prepared by homogenizer providing high speed, which helps in the formation of homogenous emulsion droplets. This technique provides higher dispersion of particles with less heterogeneity [20]. In recent advancement to this technique, a large volume of hot oil at 120°C is added drop by drop to the continuous medium to form an emulsion. This technique is based on rapid evaporation of continuous phase that converts the albumin molecules into irreversible ENPs [20].

In simple coacervation method, Sailaja, and group of researchers reported preparation of bovine serum albumin (BSA) nanoparticle. In a buffered solution of BSA, anhydrous ethyl alcohol is added slowly and stopped when solution becomes turbid. For cross-linking, 25% glutaraldehyde solution is added [18]. The non-reacted aldehyde groups are blocked with addition of ethanolamine; and to stabilize the process, tween-20 is introduced in the matrix.

Centrifugation and membrane filtration techniques can be used to obtain desired ENPs. In cross-linking w/o (water into oil emulsion) emulsion method, protein solution is homogenized under high pressure to form a water-in-oil emulsion. This w/o emulsion is added to the hot oil, which forms the suspension solution. The temperature of the suspension is maintained above 100°C and is continuously stirred, which leads to denaturation and aggregation of the protein and then water content is evaporated resulting in formation of ENPs [7].

Transmission electron microscopy (TEM), dynamic light scattering (DLS), ultraviolet (UV) spectroscopy and Fourier transform infrared (FTIR) spectroscopy are widely used tools and techniques to characterize different properties of the enzyme aggregates/ENPs.

10.2.1   SHAPE AND SIZE

The size of NPs is important for their biodistribution and bioretention in the tissues and cells and it is determined by DLS. DLS compares the Brownian motion of NPs, their velocity (translational diffusion coefficient) to figure out this to the size of NPs using stokes-Einstein equation [13]. The structural analysis of ENPs with TEM revealed that the spherical shaped monomeric/free ENPs have diameter in range of 2 to 18 nm; while aggregates have diameter in range of 100 to 200 nm.

10.2.2   COLOR

The enzyme after aggregation may either retain the same color or may have a different color. ENPs may or may not have the same color due to its free form in solution [14]. The ENPs of horseradish peroxidase (HRP) appear white in color; while protein in solution was transparent to electron beam.

10.2.3 UV ABSORPTION SPECTRA

The free enzyme absorption spectrum depends on its functional compositions. For instance, GOx UV spectrum shows two peaks: first at 235 nm due to peptide bond; and the second peak due to the presence of aromatic amino acids. The GOx nanoparticle shows absorption peak at 235 nm; while the free form of enzyme shows at 267 nm. This higher to lower wavelength is known as blue shift, which confirms the formation of ENPs [4, 21]. After aggregation to enzyme nanoparticle, the absorption peaks may decrease or increase because of the molecular interaction of protein.

10.2.4 FTIR SPECTRA

In a study of GOx enzyme nanoparticle, FTIR spectra of the enzyme nanoparticle demonstrated a band at 1673.20 cm^{-2}. This might be attributed to stretching of C=N bond after glutaraldehyde cross-linking along with vibration band of amide-I band (1,700–1,600 cm^{-1}) caused by C=O stretching vibration of peptide linkages and amide-II band (1,600–1,500 cm^{-1}) in enzyme backbone resulting from the combination of N-H bond in plane bending and C-N stretching of peptide group. It was also found that the vibration band of the functionalized ENPs was broadened to 3,000–3,500 cm^{-1} as compared to the free enzyme molecules, which confer that free amine group is introduced to GOx nanoparticle aggregate by cysteamine dihydrochloride [12].

10.3 APPLICATIONS OF ENZYME NANOPARTICLES (ENPS)

In the development of biosensors, ENPs play an essential role. Biosensor is used to detect the biological element, which is connected to a transducer that converts the detected response or signal to measurable units. The magnitude of the signal is directly proportional to the concentration of a particular chemical or biological compound. GOx enzyme-based biosensors sense the consumption of oxygen or the production of hydrogen peroxide. The biosensors are preferred because of their advantages over the other analytical devices, such as: simplicity, sensitivity, specificity, and rapidity (Table 10.1). The ENPs have been used to develop amperometric biosensors for detection of glucose, hydrogen peroxide, uric acid, and cholesterol in biological fluids [3, 12, 14, 21].

TABLE 10.1 Examples Nanoparticles Used in Different Bioprocessing Industries

Fields	Nanoparticles	Use	Remark	References
Biosensors	Glucose oxidase ENPs	To determine the glucose level in blood	Based on electrochemical interactions.	[12]
	Uricase ENPs	Uric acid determination in body		[3]
	Cholesterol oxidase ENPs	Cholesterol level determination in body		[4]
	Horseradish peroxidase ENPs	H_2O_2 conc. determination	Used in health, food, and pharma industries	[12]
Biofuel and bioenergy	CuO, ZnO, Fe, Pd, Ag, and Cu encapsulated on SiO_2	The effect of different nanoparticles is investigated for biogas production	Methane gas production by anaerobic digestion.	[1, 15]
	Fe_3O_4, NZVI, TiO_2, Pt/SiO_2, Ni/SiO_2, Co/SiO_2	—	Increase the methane and biogas production	[1, 2, 22]
Biomedical	Functionalized gold nanoparticle	Targeted drug delivery, bioimaging, gene delivery and diagnostic applications.	—	[23]
	Superparamagnetic iron oxide NPs	*In vivo* biomedical applications such as immunoassay, tissue repair, detoxification of biological fluids.	—	[8]

10.3.1 BIOSENSORS BASED ON ENZYME NANOPARTICLES (ENPS)

10.3.1.1 PEROXIDASE ENZYME NANOPARTICLE

Several oxidase enzymes are responsible for the production of *in-vivo* hydrogen peroxide that has capability to generate free hydroxyl radicals; and these free radicals can damage the proteins and lipid structures of the tissues resulting in atherosclerosis and aging. To analyze the hydrogen peroxide production, an amperometric biosensor using an immobilized HRP enzyme nanoparticle is constructed and used.

The ENPs were immobilized on the surface of gold (Au)-electrode through thiol-bond without promoters and mediators. Pt wire was an auxiliary electrode, Ag/AgCl (silver chloride) was a reference electrode and enzyme nanoparticle/Au was working electrode. These three electrodes were used for electrochemical measurements. A linear relationship between biosensor response and the hydrogen peroxide concentration was considered to evaluate the efficiency of a biosensor. The detection of hydrogen peroxide through the sensor was limited to a 0.1 μM at a "signal to noise ratio: of 3 and the coefficient of variation (CV) for hydrogen peroxide concentration was 2.54%.

10.3.1.2 GLUCOSE OXIDASE (GOX) ENZYME NANOPARTICLE

For the medical and diagnosis management, glucose level in the blood is regularly measured. There are two types of glucose biosensors for the measurement system that are based on amperometric and dissolved oxygen (DO) meter [12, 21]. This GOx enzyme-based biosensor measures the glucose concentration on the basis of utilization of DO in the blood, by the oxidation of glucose to gluconic acid and hydrogen peroxide.

An amperometric biosensor is based on GOx NPs; and it is constructed by covalent immobilization of enzyme nanoparticle onto the glutaraldehyde activated and amino-functionalized Pt electrode as a working electrode. Ag/AgCl as reference electrode is connected to working electrode through a microammeter in series. The applied potential across the electrodes is 0.4 V/s.

The DO metric glucose biosensor is constructed by immobilizing ENPs on nitrocellulose membrane pretreated with chitosan, coupled with glutaraldehyde. This designed membrane is then fitted over the sensing part of combined electrode of DO meter. In amperometric biosensor, phosphate buffer with 0.1 M and pH of 7.4 is used to immerse the electrodes. In case of DO metric biosensor, buffer with 0.05 M and pH 6 is used. The varying concentration of glucose and their responses in µAmp or DO in mg/L is measured.

10.3.1.3   URICASE ENZYME NANOPARTICLE

Uric acid in urine or blood or both is an indication of metabolic and kidney disorders. Irregular levels of uric acid in human body fluid occurs because of hyper-uricemia, gout, cardiovascular, and chronic renal disease and Lesch-Nyhan syndrome. There are various methods available to detect levels of uric acid; but uricase enzyme nanoparticle-based biosensor method is relatively rapid, simple, sensitive, specific, and user-friendly [3].

This biosensor is constructed employing uricase nanoparticle. The amperometric uric acid biosensor is designed by covalent immobilization of thiolated uricase NPs on the surface of gold wire as a nanoparticle electrode or working electrode. Ag/AgCl and Pt wire are standard and auxiliary electrodes, respectively, and these are connected to Galvanostat. The system uses uric acid in tris-buffer with varying concentrations (0.1–1.0 mM). The current generated in µAmp is measured as response in the range of 0–1.5 V/s.

10.3.1.4   CHOLESTEROL OXIDASE (CHX) ENZYME NANOPARTICLES (ENPS)

Hyper cholesterol anemia is a disease caused by abnormal level of cholesterol in the serum. Therefore, medical management and diagnosis for the detection of cholesterol is always essential. Cholesterol oxidase (CHx) enzyme nanoparticle is used to construct an amperometric biosensor, which makes the cholesterol determination [4]. This biosensor works on electrochemical reactions. The construction of the biosensor is very similar to uricase enzyme nanoparticle-based biosensor. The biosensor has the advantage of measuring cholesterol at lower concentration of 1.56 mg/dL.

10.3.2   ENZYME NANOPARTICLES (ENPS) IN BIOFUELS AND BIOENERGY

Enzymes have a special role in biofuels production, such as: biogas, bioethanol, or biodiesel. The enzymes help in hydrolyzing biomass in biofuel production. In biofuels production, nanotechnology-based NPs can help as carrier and immobilizing material, which may increase the efficiency of enzymes [13]. For more details, the reader is referred to Chapter 9 in this book volume.

The metallic NPs with higher chemical reactivity are being applied for biomass pretreatment, water treatment, soil remediation, groundwater remediation and biohydrogen production [6]. In a study, it was demonstrated that incorporation of iron NPs along with organic-materials enhanced the degradation efficiency and biogas production. However, presence of iron NPs can also conclude that, organic matter can be used as compost [16].

The difficulties generally faced in the process are less catalyst activity and low reaction rate. For the improvement of the biodiesel production process [19], different types of nano catalysts are being explored. Employing nano-magnetic solid base catalyst of $KF/CaO-Fe_3O_4$ [9], the yield of biodiesel was enhanced: (1) with KF/CaO nano catalyst from Chinese tallow seed oil [24]; (2) with heterogeneous solid base nano catalyst from soybean oil [17], and (3) magnetic nano-biocatalyst aggregates from grease and sulfated zirconia nanoparticle catalysts [5].

10.3.3   SCOPE OF ENZYME NANOPARTICLES (ENPS) IN BIOMEDICAL INDUSTRY

Owing to the size smaller than cells, NPs have several unique properties that can be exploited for biomedical application. The NPs can easily penetrate into tissues and cells diffusing through extracellular matrix. Utilizing the unique properties of NPs and nanomaterials, these could be used as carrier for enzyme molecules to favor desirable kinetics and specificity for substrate [10].

The blood serum glucose level is determined by DO metric biosensor. The biosensor determines the glucose content in the serum. Similarly, cholesterol level is measured with an amperometric cholesterol biosensor. The CHx NPs-based biosensor measures the total cholesterol level in serum.

10.4   FUTURE PROSPECTS

With the advances in nanotechnology and nanoscience, ENPs have already been used for biosensors, energy production, food industry and medicine. The nanoscale devices are being manufactured to reproduce the nanodevices existing in nature and include DNA, membrane, proteins, and other natural biomolecules. In development of enzyme nanoparticle, the major challenge is retaining the enzyme functionality for their several applications. The improvement in ENPs stability will open wide range of their use. The interaction of enzyme and NPs or nanomaterial, which is the basis of using the ENPs, could help to understand their functionality at nano level. The products based on ENPs (such as: cosmetics, bioprocessing value, biosensors, etc.), have large market in throughout the world.

10.5   SUMMARY

The ENPs are being prepared by biochemical interaction of enzyme/protein molecules and nanomaterials. The emulsification, de-solvation, and coacervation are widely used strategies in enzyme nanoparticle preparation. TEM, UV spectroscopy and FTIR spectroscopy are the tools to characterize different properties of the enzyme aggregates/ENPs, such as: size, shape, surface structure, etc. In bioprocessing of foods, drugs, and pharmaceuticals, biofuels, cosmetics, and in the development of biomolecule-based devices, nanotechnology is getting more attention because of its specific properties at nano scale.

KEYWORDS

- bioprocessing
- biosensors
- bovine serum albumin
- dynamic light scattering
- enzymes
- food science
- nanoparticles

REFERENCES

1. Al-Ahmad, A. E., Hiligsmann, S., Lambert, S., Heinrichs, B., Wannoussa, W., Tasseroul, L., Weekers, F., & Thonart, P., (2014). Effect of encapsulated nanoparticles on thermophillic anaerobic digestion. *Oral Presentation at 19th National Symposium on Appl. Biological Science* (p. 6). Gembloux: Belgium.
2. Casals, E., Barrena, R., García, A., González, E., Delgado, L., Busquets-Fité, M., Font, X., et al., (2014). Programmed iron oxide nanoparticles disintegration in anaerobic digesters boosts biogas production. *Small, 10*(14), 2801–2808.
3. Chauhan, N., Kumar, A., & Pundir, C. S., (2014). Construction of an uricase nanoparticles modified Au electrode for amperometric determination of uric acid. *Applied Biochemistry and Biotechnology, 174*(4), 1683–1694.
4. Chawla, S., Rawal, R., Sonia, Ramrati, & Pundir, C. S., (2013). Preparation of cholesterol oxidase nanoparticles and their application in amperometric determination of cholesterol. *Journal of Nanoparticle Research, 15,* 1934–1941.
5. Chen, G., Guo, C. Y., Qiao, H., Ye, M., Qiu, X., & Yue, C., (2013). Well-dispersed sulfated zirconia nanoparticles as high-efficiency catalysts for the synthesis of bis(indolyl)methanes and biodiesel. *Catalysis Communications, 41,* 70–74.
6. Demetzos, C., (2016). *Pharmaceutical Nanotechnology* (p. 215). Singapore: Springer.
7. Ezpeleta, I., Arangoa, M. A., Irache, J. M., Stainmesse, S., Chabenat, C., Popineau, Y., & Orecchioni, A. M., (1999). Preparation of ulex europaeus lectin-gliadin nanoparticle conjugates and their interaction with gastrointestinal mucus. *International Journal of Pharmaceutics, 191*(1), 25–32.
8. Gupta, A. K., & Gupta, M., (2005). Synthesis and surface engineering of iron oxide nanoparticles for biomedical applications. *Biomaterials, 26*(18), 3995–4021.
9. Hu, S., Guan, Y., Wang, Y., & Han, H., (2011). Nano-magnetic catalyst $KF/CaO–Fe_3O_4$ for biodiesel production. *Applied Energy, 88*(8), 2685–2690.
10. Kharat, M. G., Murthy, S., & Kamble, S. J., (2017). Environmental applications of nanotechnology: A review. *ADBU Journal of Engineering Technology, 6*(3), 1–11.
11. Kim, J., & Grate, J. W., (2003). Single-enzyme nanoparticles armored by a nanometer-scale organic/inorganic network. *Nano Letters, 3*(9), 1219–1222.
12. Kundu, N., Yadav, S., & Pundir, C. S., (2013). Preparation and characterization of glucose oxidase nanoparticles and their application in dissolved oxygen metric determination of serum glucose. *Journal of Nanoscience & Nanotechnology, 13*(3), 1710–1716.
13. Leonard, D., Krishnamurthy, M., Reaves, C. M., Denbaars, S. P., & Petroff, P. M., (1993). Direct formation of quantum-sized dots from uniform coherent islands of InGaAs on GaAs surfaces. *Applied Physics Letters, 63*(23), 3203–3205.
14. Liu, G., Lin, Y., Ostatná, V., & Wang, J., (2005). enzyme nanoparticles-based electronic biosensor. *Chemical Communications, 27,* 3481–3483.
15. Luna-delRisco, M., Orupõld, K., & Dubourguier, H. C., (2011). Particle-size effect of CuO and ZnO on biogas and methane production during anaerobic digestion. *Journal of Hazardous Materials, 189*(1), 603–608.
16. Mao, C., Feng, Y., Wang, X., & Ren, G., (2015). Review on research achievements of biogas from anaerobic digestion. *Renewable and Sustainable Energy Reviews, 45,* 540–555.

17. Qiu, F., Li, Y., Yang, D., Li, X., & Sun, P., (2011). Heterogeneous solid base nanocatalyst: Preparation, characterization and application in biodiesel production. *Bioresource Technology, 102*(5), 4150–4156.

18. Sailaja, A., Amareshwar, P., & Chakravarty, P., (2011). Different techniques used for the preparation of nanoparticles using natural polymers and their application. *International Journal of Pharmacy and Pharmaceutical Science, 3*(2), 45–50.

19. Sajith, V., Sobhan, C. B., & Peterson, G. P., (2010). Experimental investigations on the effects of cerium oxide nanoparticle fuel additives on biodiesel. *Advances in Mechanical Engineering, 2,* 11. Article ID: 581407.

20. Scheffel, U., Rhodes, B. A., Natarajan, T., & Wagner, H. N., (1972). Albumin microspheres for study of the reticuloendothelial system. *J. Nucl. Med, 13*(7), 498–503.

21. Sharma, S., Shrivastav, A., Gupta, N., & Srivastava, S., (2011). Amperometric biosensor: Increased sensitivity using enzyme nanoparticles. *Int. Conf. Nanotechnol. Biosens., 2,* 24–29.

22. Su, L., Shi, X., Guo, G., Zhao, A., & Zhao, Y., (2013). Stabilization of sewage sludge in the presence of nanoscale zero-valent iron (nZVI): Abatement of odor and improvement of biogas production. *Journal of Material Cycles and Waste Management, 15*(4), 461–468.

23. Tiwari, P. M., Vig, K., Dennis, V. A., & Singh, S. R., (2011). Functionalized gold nanoparticles and their biomedical applications. *Nanomaterials, 1*(1), 31–39.

24. Wen, L., Wang, Y., Lu, D., Hu, S., & Han, H., (2010). Preparation of KF/CaO nanocatalyst and its application in biodiesel production from Chinese tallow seed oil. *Fuel, 89*(9), 2267–2271.

CHAPTER 11

TECHNOLOGY OF CARBON NANOTUBES FOR WASTE WATER TREATMENT

BHARTI VERMA and CHANDRAJIT BALOMAJUMDER

ABSTRACT

Carbon nanotubes (CNTs) are used as adsorbents, catalysts or catalyst support, membranes, and electrodes to solve many environmental potential problems. Research has been directed onto the surface modifications of CNTs and their effect on its adsorption behavior and their subsequent utilization in the removal of obnoxious heavy metals from wastewater. This chapter presents a review of number of parameters and their subsequent effects, such as: pH of the solution, dosage of CNTs, contact time, temperature of the solution and the charge on the surface on the use of CNT as adsorbent for heavy metal removal. It is established that the adsorption capacity of CNTs towards heavy metals can be enhanced by surface modifications=. New methods which do not cause any harm to the environment for the modification of CNTs surface, regeneration/reuse of CNTs require investigation. Future research exploring the techniques for a cost-effective CNT production is recommended.

11.1 INTRODUCTION

Heavy metals have become an eco-toxicological hazard of prime interest [105, 154]. Their continuous release from various industries into the ecosystem creates pollution [119]. Being non-biodegradable in nature, heavy metals accumulate in living organisms and the contaminants are detrimental to human health.

The toxicological effects of heavy metals are such that they cause the blood pressure to increase, hamper with speech, tiredness, and disrupt the sleeping pattern, poor concentration, autoimmune diseases, aggressive behavior, vascular occlusion, irritability, depression, increased allergic reactions, mood swings, and memory loss, etc. They are known to interrupt with the human cellular enzymes. Several heavy metals (specifically iron, magnesium, zinc, copper) are needed in micro-quantities for the overall development of body. However, their presence in huge amounts may be threatening [150]. The maximum concentration limits (mg/L) of different heavy metals are as follows [101]:

- Arsenic: 0.05;
- Cadmium: 0.01;
- Chromium: 0.05;
- Copper: 0.25;
- Lead: 0.006;
- Mercury: 0.00003;
- Nickel: 0.20;
- Zinc: 0.80.

Being an important trace element vital for the physiological functions of the body, Zinc in small amounts is truly essential for human health. However, in large quantities, Zinc is known to cause stomach cramps, skin irritations, vomiting, nausea, and anemia, etc. [159]. Copper plays a significant role in metabolism. Nonetheless, copper in excessive quantities leads to convulsions, cramps, vomiting, etc. [20].

Nickel surpassing its maximum concentration limit is known to cause problems in kidney and lung functioning in addition to pulmonary fibrosis, gastrointestinal distress, and skin dermatitis, etc. Also, Nickel is a proven carcinogen [99]. Being a neurotoxin, mercury can harm the central nervous system (CNS) and its high concentration may cause impairment of pulmonary and kidney function, chest pain and dyspnea [16].

U.S. Environmental Protection Agency (EPA) has declared the Cadmium to be a probable human carcinogen. The symptoms of chronic exposure of cadmium are anemia, headache, dizziness, mood swings, muscle weakness, sleep disorders, and kidney damages, etc. [64]. Chromium exists in mainly two oxidation states in the environment, such as: Cr(VI) and Cr(III) [144].

The most soluble and mobile forms of Cr(VI) in aqueous environments are: $HCrO^{-4}$, CrO_2^{-4}, and $Cr_2O_2^{-7}$. Considered to be detrimental for both flora and fauna, Cr(VI) is viewed as a carcinogenic. Also, Cr(VI) is almost 300 times more toxic and subsequently harmful than Cr(III) [37]. Therefore, it becomes mandatory to treat waste water and remove these noxious heavy metal ions from the water.

11.2 SOURCES OF HEAVY METALS AND THEIR DETRIMENTAL EFFECTS

Table 11.1 shows an overview of the sources of heavy metals along with their toxic effects and the methods, which are available for their consequent removal.

11.1.1 LEAD

Being non-biodegradable, lead accumulates in the body and causes detrimental effects. Drinking water is a major source of lead ingestion into the body via digestive tract and lungs and then gets expanded across the whole body. Lead beyond its permissible limit is carcinogenic and can also cause damage to nervous system, anemia, kidney damage, etc.

11.1.2 CADMIUM

The biological half-life of cadmium in humans is around 10 to 35 years. The major industries, in which there is application of cadmium, are nickel cadmium batteries, Cadmium tellurium thin film solar cells, electroplating, etc. Beyond its permissible limit, it can lead to permanent damage to kidneys.

11.1.3 CHROMIUM

Chromium occurs in two forms with oxidation states of 3 and 6. The Cr(III) is around 300 times less toxic than Cr(VI). Therefore, it becomes evident to reduce the hexavalent form to its trivalent form by using different approaches. Hexavalent chromium (Cr(VI)) is a cause of serious diarrhea,

TABLE 11.1 Sources and Toxic Effects of Heavy Metals and Their Treatment Methods

Heavy Metals	Sources	Harmful Effects	Available Treatment Methods	References
Arsenic	• Pesticides • Fungicides • Runoff from orchards • Runoff from glass and electronics production wastes • Mining and metallurgical operations	• Skin cancers • Damage to lungs, kidneys, and causes internal tumors. • Problem in circulatory system • Death of new born babies and their sudden weight loss • Damaging the ability to hear completely • Causing injury to reproductive system	• Coagulation and flocculation • Softening by using lime • Removal using Iron and manganese • Activated alumina adsorption • Reverse osmosis • Membrane filtration	[88, 103]
Cadmium	• Steel and plastics industries • Blow down from cooling towers. • Electroplating operations. • Nickel-cadmium batteries • Nuclear emission plants • Cadmium telluride thin film solar cells • Fertilizers	• Kidney and skeletal damage • Bronchiolitis • Cancer • Chronic obstructive pulmonary disease	• Precipitation and Coagulation • Ion-exchange • Softening	[8]
Chromium	• Discharge from steel and pulp mills. • Erosion of natural deposits	• Severe diarrhea • Allergic dermatitis • Vomiting	• Membrane technologies • Electro kinetics • Coagulation and Floatation	[1, 7, 27, 53, 65, 86, 114]

TABLE 11.1 (Continued)

Heavy Metals	Sources	Harmful Effects	Available Treatment Methods	References
	• Cooling tower blowdown • Electroplating	• Pulmonary congestions; liver and kidney damage	• Photocatalytic degradation • Treatment using cyanide • Solvent extraction • Bio sorption	
Copper	• Petroleum refinery • Mining operations • Electroplating • Fungicides, algaecides, etc.	• Enhanced BP levels • Short term: Gastro-intestinal distress • Long term: liver or kidney damage • Serious toxicological concerns	• Membrane filtration processes • Electrochemical treatment • Chemical precipitation	[13, 66, 68, 71, 72, 75, 84, 119, 135]
Lead	• Paint • Pesticides • Automobile emissions • Mining • Burning of coal • Smoking	• Anemia • Insomnia • Cancer • Damage to kidneys • Impeding intelligence in children and disrupting their behavior • Damage to the nervous system	• Reverse osmosis • Adsorption process • Membrane separation processes • Chemical precipitation • Filtration • Cementation	[67]
Mercury	• Runoff from landfills and croplands • Discharge from refineries and factories	• Damage to kidneys, chromosomes • Blindness and damage to Reproductive systems	• Precipitation/co-precipitation • Lime softening • Solvent extraction	[40, 46, 102, 145, 152, 156, 160]

TABLE 11.1 *(Continued)*

Heavy Metals	Sources	Harmful Effects	Available Treatment Methods	References
	• Pesticides, Batteries, and paper industry • Chlor alkali industry wastewater	• Involuntary mobilization • Immune, hematologic, cardiovascular, respiratory system and brain	• Bioremediation and Adsorption • Ion exchange and Membrane filtration	[10, 15, 24, 34, 35, 49, 52, 54, 63, 76, 107, 127, 132, 139, 153]
Nickel	• Rechargeable batteries • Nickel cadmium batteries • Sanitary installations • Printing • Electroplating • Fossil fuel combustion	• Respiratory failure • Lung cancer, nose cancer, larynx cancer and prostate cancer • Birth defects • Asthma and chronic bronchitis • Heart disorders	• Reduction and precipitation • Adsorption • Ion exchange • Electrolysis	
Zinc	• Manufacture of brass, paints, and rubber products • Cosmetics • Wood pulp production • Galvanizing, Alloys, and metals • Thermoplastics	• Stomach nausea • Pancreas damage • Arteriosclerosis • Skin irritations • Anemia	• Precipitation using chemicals • Adsorption process • Membrane technologies • Ion exchange resins	[26, 36, 47, 59, 95]

kidney damage, and can cause cancer. The chromium in drinking water will arise if there is a contamination of industrial wastewater. The major industries by which chromium is released in the discharge are: electroplating, blowdown of cooling water, chromate preparation and leather tanning.

11.1.4 ARSENIC

The presence of arsenic is contributed by activities of human nature (such as: mining, wood preservation, metallurgical operations, pesticides, etc.). The population exposed by drinking water containing arsenic is reported to have cancer related health issues and skin cancer, internal tumors of various forms, vascular diseases, etc.

11.1.5 MERCURY

Mercury is liberated from the deposits of minerals, fossil fuels and ores and gets accumulated in the water bodies and earth surface soil. Mercury in its most elemental form itself is quite dangerous. Methyl mercury is the most noxious of all forms of mercury. Mercury can cause damage to kidneys, brain, and respiratory and cardiovascular systems. The people eating a lot of fish are most hit by the detrimental effects of mercury.

11.1.6 COPPER

Having a wide range of applications in both industrial as well as agricultural processes, copper is emitted into the surroundings in many forms. Mercury is the only heavy metal which is reported to be more toxic than copper. Copper exposure can lead to life threatening diseases such as sudden rise in blood pressure, kidney damage, liver damage, vomiting, cramps, convulsions, and even death.

11.1.7 ZINC

Zinc is an essential element in the functioning of the living tissues. However, if present beyond its permissible limit, it could be highly detrimental to the human health. The major sources of zinc into the water body discharge are the electroplating industries, brass manufacturing units, galvanizing operations, production of wood, etc. Zinc exposure can lead to irritation of skin, stomach nausea, vomiting, and anemia, etc.

11.1.8 NICKEL

Nickel is an obnoxious heavy metal, which is highly toxic in nature. There are wide range of industrial applications, which are the sources of nickel pollution in water, such as: manufacturing of batteries, alloys manufacturing, electroplating, silver refineries, etc. The detrimental effects of nickel are chest pain, asthma, cyanosis, lung cancer, nausea, etc.

11.2 AVAILABLE TECHNIQUES TO REMOVE HEAVY METALS FROM WASTEWATER

There are numerous treatment technologies for the removal of heavy metals. The technology is chosen depending on the initial concentration of heavy metals in the feed and the economics of the process. The summary of the advantages and disadvantages of different technologies is given in Table 11.2.

Every method has its own pros and cons. As an example, chemical precipitation leads to sludge generation. Membrane filtration is costly and produces large residual rejects. Although highly efficient, required regeneration of resins causing secondary pollution limits the use of ion exchange method. Flotation has limitation of high-cost operation and maintenance expenses.

One of the most economical, effective, and proven technology for wastewater treatment is adsorption [46, 104, 107, 152, 160]. The most favorable advantage of the adsorption process is that by employing desorption process, the adsorbent could be suitably regenerated [118, 133, 155].

The adsorbents for the heavy metal adsorption are: activated carbon [30, 61], sewage sludge ash [108], sugar beet pulp [48, 70], Kaolinite [42, 87], Bagasse [44, 81, 97], resin [69, 124], fly ash [18, 58, 138], chitosan [79, 131, 141], zeolite [17, 39, 57, 93]. However, there is always a quest for adsorbents with better adsorption capacities and efficiencies for heavy metal removal from wastewater and recently developed CNTs are proving to be successful. CNTs have used for removal of cadmium [128, 140, 147], Lead [19, 106], Nickel [6, 55], Cobalt [2, 82], Zinc [73, 100], Arsenic

TABLE 11.2 Pros and Cons of Using Different Techniques for Wastewater Treatment

Wastewater Treatment Technique	Pros	Cons	References
Adsorption	• Easier operation • Bind metal efficiently • Applicable in a broader range of pH • Cost effective	• Waste products are produced simultaneously • Selectivity is an issue	[13, 89, 102, 119, 126]
Biological treatment	• Eco-friendly • Requires less energy	• Slow process	[91]
Chemical precipitation	• Operation is easy • Cheaper technology	• Sludge is generated and the associated cost of its disposal • Not very efficient in dealing dilute solutions	[11, 92]
Coagulation and flocculation	• Also removes turbidity • Sludge produces is settled able. • Dewatering qualities	• Huge sludge volumes generated • Must be accompanied by another techniques • Large quantities of chemicals required	[43, 92, 115]
Electro dialysis	• Highly selective	• Operating cost is huge considering the membrane fouling and energy required.	[38, 50, 101, 122, 136]
Ion exchange	• High treatment capacity • Recovery of metal value • Less sludge volume produced • Better efficiencies of removal • Kinetic rate is fast	• High cost due to synthetic resins • Regeneration further causes secondary pollution	[25, 43, 54, 146]

TABLE 11.2 *(Continued)*

Wastewater Treatment Technique	Pros	Cons	References
Membrane filtration	• Highly selective • Space required is less • Pressure required is low	• Operating costs are higher due to membrane fouling • Permeate flux is low • Process is a little complex	[9, 11, 12, 96, 148]
Oxidation	• Electricity is not required	• Rusting	[3]
Photo catalysis	• It removes metals as well as organic pollutants • Very less toxic by products	• Applications are quite limited • Time required is long for efficient removal of heavy metals.	[4, 43]

[109, 137, 158], Copper [23, 28], Mercury [45, 112], chromium [60, 77]. The literature review indicates that CNTs are promising adsorbents for removal of heavy metals.

11.2.1 CARBON NANOTUBES (CNTS)

When cylindrical graphene sheets are rolled into tube structures, they lead to the formation of CNTs that may be single-walled (SW) or multiple-walled (MW) [113]. If there is a single shell of graphene, they are called SWCNTs (single walled carbon nanotube), and if there are multiple concentric shells of graphene, they are referred to as MWCNTs. CNTs are known to be stronger than the alkanes because of the presence of sp^2 bonds unlike the sp^3 bonding present in alkanes.

CNTs are widely used in the wastewater treatment due to their huge surface area, high porosity, hollow structure, and strong interaction between the contaminant and the CNTs [143]. It has been reported that heavy metals [128], small molecules [83], organic contaminants [110], and nucleotides [116] are adsorbed effectively on the CNTs [51]. It could be seen that there are four sites, namely: (a) internal sites, which are hollow interior of CNTs that could only accessed by opening the open ends; (b) interstitial sites (ICs); (c) grooves; and (d) outside surface.

The adsorption takes place at the interstitial sites and the outer surface and the grooves on [121]. Also, unblocking of the ends of CNTs leads to increase in the number of adsorption sites, so that saturation capacity is also increased [85, 123]. The maximum adsorption capacity was derived by using the Langmuir adsorption isotherm (Table 11.3). It can also be deduced from Table 11.3 that the CNTs which have undergone acid treatment prior to adsorption experiment showed much higher capacities than the raw CNTs. This could be due to the electrostatic interaction between the CNT's surface charge and the charged divalent heavy metal ions. Table 11.3 indicates maximum adsorption capacities of CNTs for different heavy metals under different conditions.

11.2.2 SYNTHESIS OF CNTS

Presently, four methods are used for the synthesis of CNTs, namely: arc discharge method [22], gas phase catalytic growth from monoxide of

TABLE 11.3   Literature Review for Heavy Metal Removal by CNTs

Adsorbent	Targeted Heavy Metal	Maximum Adsorption Capacity (Q_m)	Optimized Parameters	Interaction Mechanism	Isotherm Model Used	References
Acidified multi walled CNTs (MWCNTs)	Ni^{2+}	17.86 mg/g	Initial concentration of Ni^{2+} = 20 mg/L; at 65°C	Ion exchange	Langmuir	[107]
Activated carbon coated with CNTs	Cr^{6+}	9 mg/g	Initial Cr(VI) concentration 0.5 mg/L; adsorbent dosage 40 g/L; pH 2; 60 min; 100 rpm	NA	Langmuir	[65]
Carboxyl group attached single walled CNTs	Cd^{2+}	55.89 mg/g	Initial Cd(II) concentration 20 mg/L; pH 5; adsorbent dosage 50 mg/L; 298 K	Chemisorption as well as physisorption	–	[149]
Carboxylic group-MWCNT	Hg^{2+}	127.6 mg/g	Initial Hg(II) concentration 4 mg/L; pH 4	Chemical adsorption	Langmuir	[31]
CNTs/calcium alginate	Cu^{2+}	72.99 mg/g	pH 5; dosage 0.5 g/L; temperature 20°C	Specific adsorption sites; Cation exchange	Langmuir	[130]
CNTs (HNO_3)/methane-Ni	Pb^{2+}	82.6 mg/g	pH 5; dosage 0.02 g/100 mL; initial concentration of Pb(II) 10 mg/L; 6 hours	NA	NA	[120]
Diethylenetriamine MWCNTs	Pb^{2+}	58.26 mg/g;	Initial concentration 5 mg/L; 298 K; pH 6.2; dosage 100 mg/L	Both physisorption and chemisorption	Langmuir	[14]
Ethylenediamine functionalized MWCNTs	Cd^{2+}	25.7 mg/g	Initial Cd(II) concentration 5 mg/L; dosage 100 mg/L; 298 K; pH 8–9	Electrostatic attraction	Langmuir	[29]

TABLE 11.3 *(Continued)*

Adsorbent	Targeted Heavy Metal	Maximum Adsorption Capacity (Q_m)	Optimized Parameters	Interaction Mechanism	Isotherm Model Used	References
H_2SO_4/$KMnO_4$ modified CNTs	Cu^{2+}	42.7 mg/g	Initial Cr(II) concentration 43 mg/L; dosage 0.5 g/L; 320 K; 24 hours	Physisorption	Langmuir	[62]
Magnetic MWCNTs	Cr^{6+}	16.234 mg/g	pH 3; initial Cr(VI) concentration 25 mg/L; 343 K	Electrostatic interaction	Langmuir	[37]
Manganese Oxide coated CNTs	Pb^{2+}	78.74 mg/g	pH 7; initial concentration of Pb(II) 10 mg/L; 298 K	Electrostatic interaction and surface complexation	Langmuir	[94]
Multiwall carbon nanotube-zirconia nanohybrid (MWCNT-ZrO_2)	As^{3+}	2,000 µg/g	Initial As(III) concentration 100 µg/L; pH 6; 298 K	Chemisorption as well as physisorption	Langmuir Freundlich	[65]
MWCNTs purified with sodium hypochlorite	Zn^{2+}	15.77 mg/g	Initial concentration of Zn(II) 60 mg/L;	Electrostatic interactions	Langmuir Freundlich	[100]
Oxidized CNT sheets	Co^{2+}	85.7 mg/g	Initial Cr(II) concentration 1,200 mg/L; 298 K; pH 7	Chemical adsorption	Langmuir	[125]
Raw CNTs	Cr^{6+}	20.56 mg/g	Initial Cr(VI) concentration 33.28 mg/L; pH 7.5	Anion exchange	Langmuir	[33]

carbon [78], laser ablation [98] and chemical vapor deposition (CVD) from hydrocarbons (HCs) [157].

In CVD, thermal decomposition of HC in the presence of a metal catalyst takes place. This method is quite simple, easier to operate and cost-effective at a lower temperature and ambient pressure. Owing to the versatility of CVD, for example any state of HC (solid, liquid or gas) could be utilized for thermal decomposition. It enables the use of a wide variety of substrates and CNTs can be produced in different forms, such as: powder, aligned or entangled, straight or coiled, thick or thin films.

11.2.3   STRUCTURE AND PROPERTIES OF CNTS

SWCNTs can be described by a single vector, $\vec{C}$, which is commonly known as the chiral vector. When two atoms in a planar graphene sheet are selected, chiral vector points from one atom to another atom and could be defined by the following equation:

$$\vec{C} = n\vec{a}_1 + m\vec{a}_2 \qquad (1)$$

where; n and m are number of steps along the unit vectors; and $\vec{a}_1$ and $\vec{a}_2$ are the unit cell vectors of 2-D lattice formed by the graphene sheets. The axis of nanotube is at a right angle to the chiral vector [41].

Depending on the rolling angle of graphene sheets, the CNTs can have three different chiralities (armchair, zigzag or chiral). By observing the (n, m) naming procedure, the orientation of carbon atoms surrounding the structure of nanotube are identified:

If n = m, it is said to be armchair, and
Zigzag orientation is defined for m = 0.
For all other cases, the orientation is chiral.

The transport properties especially the electronic properties are dependent on the chirality of the CNTs. For example, if (2n + m) is a multiple of 3, CNTs are metallic in nature, otherwise, they are semiconducting.

For MWCNTs, whisch are a collection of SWCNTs, the identification of electronic properties is quite difficult as the chirality is different for each SWCNTs. As far as the mechanical properties are considered, experimental results have shown that they have higher mechanical properties. The Young's Modulus is about 1.2 TPa and the tensile strength ranges from 50 to 200 GPa [151].

In addition to excellent mechanical and electronic properties, they possess exceptional physical properties, which make them ideal for utilization in wide range of applications, such as: field emission, air, and water filtration, structural materials, as energy storage devices, catalyst supports, and biological treatment, etc. [5].

The thermal conductivity of SWNTs was calculated to be 6,600 W/m.K, which is more than that of a diamond by a factor of 2. On an average, the thermal conductivity of CNTs is above 1,000 W/m.K, which is much more than 430 W/m.K for silver, copper, and gold. Light adsorption, photoluminescence, and Raman scattering are related to the optical properties of CNTs.

11.2.4   SURFACE MODIFICATION OF CNTS

There are number of techniques by which surface of CNTs could be modified, such as: prior treatment with acids (HNO_3, H_2SO_4, HCl) [129], metal oxide impregnations [117], and grafting the functional molecules [32]. These techniques can modify the surface properties of raw CNTs, such as: surface area and charge, water repelling/ loving properties, and dispersion.

There are various methods applied for grafting of CNTs with different functional group to enhance their adsorption properties, such as: plasma technique, microwave technique and the chemical modification [54].

11.2.5   EFFECT OF VARIOUS PARAMETERS ON ADSORPTION OF THE HEAVY METALS ONTO CNTS

11.2.5.1   EFFECT OF PH

When the pH of solution is higher than the point of zero charge, there are electrostatic interactions among the cations and the negative surface charge. Therefore, cation adsorption is higher at pH > than the point of zero charge. Also, when the pH of solution is less than the point of zero charge, then the surface negative charge is neutralized, leading to a decrease in the adsorption of cations. The adsorption of Cu^{2+}, Pb^{2+}, Cd^{2+}, Co^{2+}, Cr^{6+}, and Ni^{2+} is observed to be maximum at pH values of: pH = 5 [21, 23], pH = 7 [111], pH = 6.8 [134], pH = 3 [74], and pH = 5.4 [34], respectively. The point of zero charge of raw CNTs ranges from 4 to 6; whereas the acid modified CNTs or the oxidized CNTs have lower value of point of zero charge [90].

11.2.5.2   EFFECT OF IONIC STRENGTH

The increase in ionic strength has a direct negative effect on the sorption of heavy metal ions. The decrease is attributed to mainly two factors, such as: (a) the activity coefficient of heavy metal ions is affected by the ionic strength and therefore there is a limitation on the transfer of ions onto the surface; and (b) At higher ion concentration, there is a formation of electrical double layer, which also limits the adsorption.

11.2.5.3   EFFECT OF CONTACT TIME

The time for which there is a contact between the adsorbent and the adsorbate is also quite important. Until equilibrium is reached, the removal of heavy metals is considerably increased as the time of contact is increased [142]. The equilibrium is further dependent on the initial metal concentration. For example, when the initial metal concentration is low, the equilibrium is achieved early, whereas it is achieved late when the initial concentration of metals is low.

11.2.5.4   EFFECT OF CNTS DOSAGE

CNTs dosage affects the sorption, but is a highly variable parameter. The dosage of CNTs can either enhance or reduce the adsorption of heavy metals ions onto the CNTs surface. The adsorption of Ni^{2+} was observed to be enhanced by enhancing the dosage of CNTs [155]. It is also evident from the fact that there are more adsorption sites when the dosage is increased. It is also seen that the dosage of acid modified CNTs required for the same adsorption is less due to better electrostatic interaction.

11.2.5.5   EFFECT OF TEMPERATURE

The sorption of divalent heavy metals ions increases on increasing the temperature. For example, the sorption of Zn^{2+} is an endothermic process and it increases on increasing the temperature [118]. This can be easily understood by the phenomenon of reduced viscosity upon increasing the temperature. The reduced viscosity allows faster diffusion of ions and therefore the sorption is increased. Also, lower value of activation energy is an indication of a faster sorption process. The value of activation however depends on the initial metal concentration. The value of activation energy is observed to be lower for higher initial concentrations of metal ions.

11.2.5.6   THERMODYNAMIC PARAMETERS

The nature of adsorption is affected by the value of enthalpy, entropy, and Gibbs free energy. The sorption of cadmium, copper, and lead was observed to be physical with respect to SWCNTs and chemical with respect to SWCNTs-COOH. Also, the sorption of various divalent heavy metals ions (such as: Lead, nickel, cadmium, copper, and Zinc) is endothermic in nature, which is confirmed by the calculated positive values of the enthalpy [90]. The positive value of entropy reveals the increase in randomness due to ongoing exchange of ions between the functional groups of CNTs and divalent metal ions. The sorption of some divalent heavy metal ions is spontaneous due to their negative Gibbs free energy [56].

11.2.6   CNTS REGENERATION

For a cost-effective adsorption process, it is beneficial to desorb the heavy metal ions and regenerate the CNTs for repeated use. The CNTs regeneration is a highly pH dependent process. The desorption of Zn^{2+} and Pb^{2+} is effective at low pH. In addition to pH, chemical treatment is another way to regenerate the adsorbent [80, 100].

For example, the MWCNTs (multi-walled carbon nanotube) are regenerated when treated with HNO_3 and Ni^{2+}, Cd^{2+} ions are desorbed. However, research is needed to find effective methods of regeneration of CNTs to optimize the economics of the process.

11.3   CHALLENGES IN USING CNTS AS ADSORBENTS

The application of CNTs as a potential adsorbent is limited by its high cost. Researchers are now in a quest of developing economical methods of large-scale production of CNTs for application in commercial operations. Also surface modifications of CNTs are practiced for better adsorption capability of CNTs. The CNTs can aggregate in a solution due to presence of Van der Waal forces. Due to these aggregations, adsorption sites are limited, which can reduce the efficiency of adsorption process.

Surface modifications may utilize huge amount of chemicals, which pose another environmental threat. Hence, there is a need to develop eco-friendly techniques of surface modifications of CNTs. Post adsorption process, there is a challenge of separating CNTs from the aqueous solution. The methods of ultra-centrifugation and membrane filtration are widely used however are limited by the high energy requirement and the costs. By far, magnetic separation to isolate CNTs is the best technique. Further efforts are needed to explore other separation techniques, which may be cost effective and efficient.

11.4 SUMMARY

The present chapter reviews all available techniques for the removal of heavy metal ions from the solution along with their advantages and disadvantages. CNTs are proving to be a promising adsorbent for the heavy metal adsorption, due to their large surface area. The acid modified, grafted functional groups CNTs have been reviewed for their selective heavy metal removal. Regeneration of CNTs is employed to make the process more economical, which is a highly pH dependent. Chemicals are also employed in the regeneration process, which poses threat to the environment.

KEYWORDS

- adsorption
- carbon nanotubes
- catalysts
- composites
- heavy metals
- surface modifications
- wastewater treatment

REFERENCES

1. Abdel-Ghani, N. T., El-Chaghaby, G. A., & Helal, F. S., (2015). Individual and competitive adsorption of phenol and nickel onto multiwalled carbon nanotubes. *Journal of Advanced Research, 6*(3), 405–415.
2. Abdel, S. O. E., Reiad, N. A., & ElShafei, M. M., (2011). A study of the removal characteristics of heavy metals from wastewater by low-cost adsorbents. *Journal of Advanced Research, 2*(4), 297–303.
3. Agnihotri, S., Mota, J. P. B., Rostam-Abadi, M., & Rood, M. J., (2006). Theoretical and experimental investigation of morphology and temperature effects on adsorption of organic vapors in single-walled carbon nanotubes. *Journal of Physical Chemistry B, 110*(15), 7640–7647.
4. Ahmaruzzaman, M., (2009). Role of fly ash in the removal of organic pollutants from wastewater. *Energy & Fuels, 23*(3, 4), 1494–1511.
5. Ahmed, M. J. K., & Ahmaruzzaman, M., (2016). Review on potential usage of industrial waste materials for binding heavy metal ions from aqueous solutions. *Journal of Water Process Engineering, 10*, 39–47.
6. Ajmal, M., Rao, R. A. K., Anwar, S., Ahmad, J., & Ahmad, R., (2003). Adsorption studies on rice husk: Removal and recovery of Cd(II) from wastewater. *Bioresource Technology, 86*(2), 147–149.
7. Al-Khaldi, F. A., Abu-Sharkh, B., Abulkibash, A. M., & Atieh, M. A., (2015). Cadmium removal by activated carbon, carbon nanotubes, carbon nanofibers, and carbon fly ash: A comparative study. *Desalination and Water Treatment, 53*(5), 1417–1429.
8. Al-Rashdi, B. A. M., Johnson, D. J., & Hilal, N., (2013). Removal of heavy metal ions by nanofiltration. *Desalination, 315*, 2–17.
9. Al-Rashdi, B. A. M., Somerfield, C., & Hilal, N., (2011). Heavy metals removal using adsorption and nanofiltration techniques. *Separation and Purification Reviews, 40*(3), 209–259.
10. Alinnor, I. J., (2007). Adsorption of heavy metal ions from aqueous solution by fly ash. *Fuel, 86*(5, 6), 853–857.
11. AlOmar, M. K., Alsaadi, M. A., Hayyan, M., Akib, S., & Hashim, M. A., (2016). Functionalization of CNTs surface with phosphonium based deep eutectic solvents for arsenic removal from water. *Applied Surface Science, 389*, 216–226.
12. Alyuz, B., & Veli, S., (2009). Kinetics and equilibrium studies for the removal of nickel and zinc from aqueous solutions by ion exchange resins. *Journal of Hazardous Materials, 167*(1–3), 482–488.
13. An, H. K., Park, B. Y., & Kim, D. S., (2001). Crab shell for the removal of heavy metals from aqueous solution. *Water Research, 35*(15), 3551–3556.
14. Ando, Y., Zhao, X. L., Sugai, T., & Kumar, M., (2004). Growing carbon nanotubes. *Materials Today, 7*(10), 22–29.
15. Atieh, M. A., (2011). Removal of chromium(VI) from polluted water using carbon nanotubes supported with activated carbon. *Procedia Environmental Sciences, 4*, 281–293.
16. Awual, M. R., (2015). A novel facial composite adsorbent for enhanced copper(II) detection and removal from wastewater. *Chemical Engineering Journal, 266*, 368–375.

17. Awual, M. R., Ismael, M., Yaita, T., El-Safty, S. A., Shiwaku, H., Okamoto, Y., & Suzuki, S., (2013). Trace copper(II) ions detection and removal from water using novel ligand modified composite adsorbent. *Chemical Engineering Journal, 222,* 67–76.

18. Awual, M. R., Yaita, T., El-Safty, S. A., Shiwaku, H., Suzuki, S., & Okamoto, Y., (2013). Copper(II) ions capturing from water using ligand modified a new type mesoporous adsorbent. *Chemical Engineering Journal, 221,* 322–330.

19. Babaa, M. R., Stepanek, I., Masenelli-Varlot, K., Dupont-Pavlovsky, N., McRae, E., & Bernier, P., (2003). Opening of single-walled carbon nanotubes: Evidence given by krypton and xenon adsorption. *Surface Science, 531*(1), 86–92.

20. Babel, S., & Kurniawan, T. A., (2003). Low-cost adsorbents for heavy metals uptake from contaminated water: A review. *Journal of Hazardous Materials, 97*(1–3), 219–243.

21. Barakat, M. A., (2011). New trends in removing heavy metals from industrial waste-water. *Arabian Journal of Chemistry, 4*(4), 361–377.

22. Bartczak, P., Norman, M., Klapiszewski, L., Karwanska, N., Kawalec, M., Baczynska, M., Wysokowski, M., et al., (2018). Removal of nickel(II) and lead(II) ions from aqueous solution using peat as a low-cost adsorbent: A kinetic and equilibrium study. *Arabian Journal of Chemistry, 11*(8), 1209–1222.

23. Belin, T., & Epron, F., (2005). Characterization methods of carbon nanotubes: A review. *Materials Science and Engineering B-Solid State Materials for Advanced Technology, 119*(2), 105–118.

24. Bhattacharyya, K. G., & Sen, G. S., (2008). Adsorption of a few heavy metals on natural and modified kaolinite and montmorillonite: A review. *Advances in Colloid and Interface Science, 140*(2), 114–131.

25. Bissen, M., & Frimmel, F. H., (2003). Arsenic - a review. Part II: Oxidation of arsenic and its removal in water treatment. *Acta Hydrochimica Et Hydrobiologica, 31*(2), 97–107.

26. Bonzongo, J. C. J., Heim, K. J., Chen, Y. A., Lyons, W. B., Warwick, J. J., Miller, G. C., & Lechler, P. J., (1996). Mercury pathways in the Carson river-Lahontan reservoir system, Nevada, USA. *Environmental Toxicology and Chemistry, 15*(5), 677–683.

27. Chang, Q., & Wang, G., (2007). Study on the macromolecular coagulant PEX which traps heavy metals. *Chemical Engineering Science, 62*(17), 4636–4643.

28. Chen, H., Li, J. X., Shao, D. D., Ren, X. M., & Wang, X. K., (2012). Poly(acrylic acid) grafted multiwall carbon nanotubes by plasma techniques for Co(II) removal from aqueous solution. *Chemical Engineering Journal, 210,* 475–481.

29. Chen, P. H., Hsu, C. F., Tsai, D. D. W., Lu, Y. M., & Huang, W. J., (2014). Adsorption of mercury from water by modified multi-walled carbon nanotubes: Adsorption behavior and interference resistance by coexisting anions. *Environmental Technology, 35*(15), 1935–1944.

30. De Gisi, S., Lofrano, G., Grassi, M., & Notarnicola, M., (2016). Characteristics and adsorption capacities of low-cost sorbents for wastewater treatment: A review. *Sustainable Materials and Technologies, 9,* 10–40.

31. Deliyanni, E. A., Peleka, E. N., & Matis, K. A., (2007). Removal of zinc ion from water by sorption onto iron-based nanoadsorbent. *Journal of Hazardous Materials, 141*(1), 176–184.

32. Depci, T., Kul, A. R., & Onal, Y., (2012). Competitive adsorption of lead and zinc from aqueous solution on activated carbon prepared from Van apple pulp: Study in single- and multi-solute systems. *Chemical Engineering Journal, 200*, 224–236.

33. Di Natale, F., Erto, A., Lancia, A., & Musmarra, D., (2011). Mercury adsorption on granular activated carbon in aqueous solutions containing nitrates and chlorides. *Journal of Hazardous Materials, 192*(3), 1842–1850.

34. Di, Z. C., Ding, J., Peng, X. J., Li, Y. H., Luan, Z. K., & Liang, J., (2006). Chromium adsorption by aligned carbon nanotubes supported ceria nanoparticles. *Chemosphere, 62*(5), 861–865.

35. Di, Z. C., Li, Y. H., Luan, Z. K., & Liang, J., (2004). Adsorption of chromium(VI) ions from water by carbon nanotubes. *Adsorption Science & Technology, 22*(6), 467–474.

36. Erdem, E., Karapinar, N., & Donat, R., (2004). The removal of heavy metal cations by natural zeolites. *Journal of Colloid and Interface Science, 280*(2), 309–314.

37. Fiol, N., Villaescusa, I., Martinez, M., Miralles, N., Poch, J., & Serarols, J., (2006). Sorption of Pb(II), Ni(II), Cu(II) and Cd(II) from aqueous solution by olive stone waste. *Separation and Purification Technology, 50*(1), 132–140.

38. Franco, P. E., Veit, M. T., Borba, C. E., Goncalves, G. D., Fagundes-Klen, M. R., Bergamasco, R., Da Silva, E. A., & Suzaki, P. Y. R., (2013). Nickel(II) and zinc(II) removal using Amberlite IR-120 resin: Ion exchange equilibrium and kinetics. *Chemical Engineering Journal, 221*, 426–435.

39. Fu, F. L., & Wang, Q., (2011). Removal of heavy metal ions from wastewaters: A review. *Journal of Environmental Management, 92*(3), 407–418.

40. Ge, Y. Y., Li, Z. L., Xiao, D., Xiong, P., & Ye, N., (2014). Sulfonated multi-walled carbon nanotubes for the removal of copper (II) from aqueous solutions. *Journal of Industrial and Engineering Chemistry, 20*(4), 1765–1771.

41. Gherasim, C. V., Krivcik, J., & Mikulasek, P., (2014). Investigation of batch electrodialysis process for removal of lead ions from aqueous solutions. *Chemical Engineering Journal, 256*, 324–334.

42. Gherasim, C. V., & Mikulasek, P., (2014). Influence of operating variables on the removal of heavy metal ions from aqueous solutions by nanofiltration. *Desalination, 343*, 67–74.

43. Goering, J., Kadossov, E., & Burghaus, U., (2008). Adsorption kinetics of alcohols on single-wall carbon nanotubes: An ultrahigh vacuum surface chemistry study. *Journal of Physical Chemistry C, 112*(27), 10114–10124.

44. Goyal, M., Bhagat, M., & Dhawan, R., (2009). Removal of mercury from water by fixed bed activated carbon columns. *Journal of Hazardous Materials, 171*(1–3), 1009–1015.

45. Gupta, V. K., Moradi, O., Tyagi, I., Agarwal, S., Sadegh, H., Shahryari-Ghoshekandi, R., Makhlouf, A. S. H., et al., (2016). Study on the removal of heavy metal ions from industry waste by carbon nanotubes: Effect of the surface modification: A review. *Critical Reviews in Environmental Science and Technology, 46*(2), 93–118.

46. Gupta, V. K., & Rastogi, A., (2008). Sorption and desorption studies of chromium(VI) from nonviable cyanobacterium nostoc muscorum biomass. *Journal of Hazardous Materials, 154*(1–3), 347–354.

47. Hadavifar, M., Bahramifar, N., Younesi, H., Rastakhiz, M., Li, Q., Yu, J., & Eftekhari, E., (2016). Removal of mercury(II) and cadmium(II) ions from synthetic wastewater

by a newly synthesized amino and thiolated multi-walled carbon nanotubes. *Journal of the Taiwan Institute of Chemical Engineers, 67*, 397–405.

48. Hadi, P., To, M. H., Hui, C. W., Lin, C. S. K., & McKay, G., (2015). Aqueous mercury adsorption by activated carbons. *Water Research, 73*, 37–55.

49. Harvey, C. F., Swartz, C. H., Badruzzaman, A. B. M., Keon-Blute, N., Yu, W., Ali, M. A., Jay, J., et al., (2002). Arsenic mobility and groundwater extraction in Bangladesh. *Science, 298*(5598), 1602–1606.

50. Hasar, H., (2003). Adsorption of nickel(II) from aqueous solution onto activated carbon prepared from almond husk. *Journal of Hazardous Materials, 97*(1–3), 49–57.

51. Hawari, A. H., & Mulligan, C. N., (2006). Biosorption of lead(II), cadmium(II), copper(II) and nickel(II) by anaerobic granular biomass. *Bioresource Technology, 97*(4), 692–700.

52. Hayati, B., Maleki, A., Najafi, F., Daraei, H., Gharibi, F., & McKay, G., (2017). Super high removal capacities of heavy metals (Pb^{2+} and Cu^{2+}) using CNT dendrimer. *Journal of Hazardous Materials, 336*, 146–157.

53. Homagai, P. L., Ghimire, K. N., & Inoue, K., (2010). Adsorption behavior of heavy metals onto chemically modified sugarcane bagasse. *Bioresource Technology, 101*(6), 2067–2069.

54. Huang, Z. N., Wang, X. L., & Yang, D. S., (2015). Adsorption of Cr(VI) in wastewater using magnetic multi-wall carbon nanotubes. *Water Science and Engineering, 8*(3), 226–232.

55. Hui, K. S., Chao, C. Y. H., & Kot, S. C., (2005). Removal of mixed heavy metal ions in wastewater by zeolite 4A and residual products from recycled coal fly ash. *Journal of Hazardous Materials, 127*(1–3), 89–101.

56. Ibrahim, H. S., Jamil, T. S., & Hegazy, E. Z., (2010). Application of zeolite prepared from Egyptian kaolin for the removal of heavy metals: II. Isotherm models. *Journal of Hazardous Materials, 182*(1–3), 842–847.

57. Ihsanullah, Abbas, A., Al-Amer, A. M., Laoui, T., Al-Marri, M. J., Nasser, M. S., Khraisheh, M., & Atieh, M. A., (2016). Heavy metal removal from aqueous solution by advanced carbon nanotubes: Critical review of adsorption applications. *Separation and Purification Technology, 157*, 141–161.

58. Ihsanullah, Al-Khaldi, F. A., Abu-Sharkh, B., Abulkibash, A. M., Qureshi, M. I., Laoui, T., & Atieh, M. A., (2016). Effect of acid modification on adsorption of hexavalent chromium (Cr(VI)) from aqueous solution by activated carbon and carbon nanotubes. *Desalination and Water Treatment, 57*(16), 7232–7244.

59. Ihsanullah, Al-Khaldi, F. A., Abusharkh, B., Khaled, M., Atieh, M. A., Nasser, M. S., Iaoui, T., et al., (2015). Adsorptive removal of cadmium(II) ions from liquid phase using acid modified carbon-based adsorbents. *Journal of Molecular Liquids, 204*, 255–263.

60. Iijima, S., (1991). Helical microtubules of graphitic carbon. *Nature, 354*(6348), 56–58.

61. Inglezakis, V. J., Stylianou, M. A., Gkantzou, D., & Loizidou, M. D., (2007). Removal of Pb(II) from aqueous solutions by using clinoptilolite and bentonite as adsorbents. *Desalination, 210*(1–3), 248–256.

62. Jiang, M. Q., Jin, X. Y., Lu, X. Q., & Chen, Z. L., (2010). Adsorption of Pb(II), Cd(II), Ni(II) and Cu(II) onto natural kaolinite clay. *Desalination, 252*(1–3), 33–39.

63. Jing, X. S., Liu, F. Q., Yang, X., Ling, P. P., Li, L. J., Long, C., & Li, A. M., (2009). Adsorption performances and mechanisms of the newly synthesized N,N' -di

(carboxymethyl) dithiocarbamate chelating resin toward divalent heavy metal ions from aqueous media. *Journal of Hazardous Materials, 167*(1–3), 589–596.

64. Jobby, R., Jha, P., Yadav, A. K., & Desai, N., (2018). Biosorption and biotransformation of hexavalent chromium [Cr(VI)]: A comprehensive review. *Chemosphere, 207,* 255–266.

65. Journet, C., Maser, W. K., Bernier, P., Loiseau, A., DelaChapelle, M. L., Lefrant, S., Deniard, P., et al., (1997). Large-scale production of single-walled carbon nanotubes by the electric-arc technique. *Nature, 388*(6644), 756–758.

66. Jung, C., Heo, J., Han, J., Her, N., Lee, S. J., Oh, J., Ryu, J., & Yoon, Y., (2013). Hexavalent chromium removal by various adsorbents: Powdered activated carbon, chitosan, and single/multi-walled carbon nanotubes. *Separation and Purification Technology, 106,* 63–71.

67. Justi, K. C., Favere, V. T., Laranjeira, M. C. M., Neves, A., & Peralta, R. A., (2005). Kinetics and equilibrium adsorption of Cu(II), Cd(II), and Ni(II) ions by chitosan functionalized with 2[-bis-(pyridylmethyl)aminomethyl]-4-methyl-6-formylphenol. *Journal of Colloid and Interface Science, 291*(2), 369–374.

68. Kabbashi, N. A., Atieh, M. A., Al-Mamun, A., Mirghami, M. E. S., Alam, M. D. Z., & Yahya, N., (2009). Kinetic adsorption of application of carbon nanotubes for Pb(II) removal from aqueous solution. *Journal of Environmental Sciences, 21*(4), 539–544.

69. Kadirvelu, K., Faur-Brasquet, C., & Le Cloirec, P., (2000). Removal of Cu(II), Pb(II), and Ni(II) by adsorption onto activated carbon cloths. *Langmuir, 16*(22), 8404–8409.

70. Kajitvichyanukul, P., Ananpattarachai, J., & Pongpom, S., (2005). Sol-gel preparation and properties study of TiO_2 thin film for photocatalytic reduction of chromium(VI) in photocatalysis process. *Science and Technology of Advanced Materials, 6*(3, 4), 352–358.

71. Kandah, M. I., & Meunier, J. L., (2007). Removal of nickel ions from water by multi-walled carbon nanotubes. *Journal of Hazardous Materials, 146*(1, 2), 283–288.

72. Kang, S. Y., Lee, J. U., Moon, S. H., & Kim, K. W., (2004). Competitive adsorption characteristics of Co^{2+}, Ni^{2+}, and Cr^{3+} by IRN-77 cation exchange resin in synthesized wastewater. *Chemosphere, 56*(2), 141–147.

73. Karnitz, O., Gurgel, L. V. A., De Melo, J. C. P., Botaro, V. R., Melo, T. M. S., Gil, R. P. D. F., & Gil, L. F., (2007). Adsorption of heavy metal ion from aqueous single metal solution by chemically modified sugarcane bagasse. *Bioresource Technology, 98*(6), 1291–1297.

74. Khosravi, J., & Alamdari, A., (2009). Copper removal from oil-field brine by coprecipitation. *Journal of Hazardous Materials, 166*(2, 3), 695–700.

75. Kuo, C. Y., (2009). Water purification of removal aqueous copper(II) by as-grown and modified multi-walled carbon nanotubes. *Desalination, 249*(2), 781–785.

76. Kurniawan, T. A., Chan, G. Y. S., Lo, W. H., & Babel, S., (2006). Physicochemical treatment techniques for wastewater laden with heavy metals. *Chemical Engineering Journal, 118*(1, 2), 83–98.

77. LaBrosse, M. R., Shi, W., & Johnson, J. K., (2008). Adsorption of gases in carbon nanotubes: Are defect interstitial sites important? *Langmuir, 24*(17), 9430–9439.

78. Lauwerys, R. R., Buchet, J. P., & Roels, H., (1979). Determination of trace levels of arsenic in human biological-materials. *Archives of Toxicology, 41*(4), 239–247.

79. Li, Y. H., Di, Z. C., Ding, J., Wu, D. H., Luan, Z. K., & Zhu, Y. Q., (2005). Adsorption thermodynamic, kinetic and desorption studies of Pb²⁺ on carbon nanotubes. *Water Research, 39*(4), 605–609.

80. Li, Y. H., Liu, F. Q., Xia, B., Du, Q. J., Zhang, P., Wang, D. C., Wang, Z. H., & Xia, Y. Z., (2010). Removal of copper from aqueous solution by carbon nanotube/calcium alginate composites. *Journal of Hazardous Materials, 177*(1–3), 876–880.

81. Li, Y. H., Wang, S. G., Luan, Z. K., Ding, J., Xu, C. L., & Wu, D. H., (2003). Adsorption of cadmium(II) from aqueous solution by surface oxidized carbon nanotubes. *Carbon, 41*(5), 1057–1062.

82. Li, Y. H., Zhu, Y. Q., Zhao, Y. M., Wu, D. H., & Luan, Z. K., (2006). Different morphologies of carbon nanotubes effect on the lead removal from aqueous solution. *Diamond and Related Materials, 15*(1), 90–94.

83. Lim, H. K., Teng, T. T., Ibrahim, M. H., Ahmad, A., & Chee, H. T., (2012). Adsorption and removal of zinc (II) from aqueous solution using powdered fish bones. *APCBEE Procedia, 1*, 96–102.

84. Liu, C. K., Bai, R. B., Hong, L., & Liu, T., (2010). Functionalization of adsorbent with different aliphatic polyamines for heavy metal ion removal: Characteristics and performance. *Journal of Colloid and Interface Science, 345*(2), 454–460.

85. Liu, C. K., Bai, R. B., & Ly, Q. S., (2008). Selective removal of copper and lead ions by diethylenetriamine - functionalized adsorbent: Behavior and mechanism. *Water Research, 42*(6, 7), 1511–1522.

86. Liu, S. S., Chen, Y. Z., De Zhang, L., Hua, G. M., Xu, W., Li, N., & Zhang, Y., (2011). Enhanced removal of trace Cr(VI) ions from aqueous solution by titanium oxide-Ag composite adsorbents. *Journal of Hazardous Materials, 190*(1–3), 723–728.

87. Liu, X. W., Hu, Q. Y., Fang, Z., Zhang, X. J., & Zhang, B. B., (2009). Magnetic chitosan nanocomposites: A useful recyclable tool for heavy metal ion removal. *Langmuir, 25*(1), 3–8.

88. Lu, C. S., Chiu, H., & Liu, C. T., (2006). Removal of zinc(II) from aqueous solution by purified carbon nanotubes: Kinetics and equilibrium studies. *Industrial & Engineering Chemistry Research, 45*(8), 2850–2855.

89. Lu, C. Y., & Chiu, H. S., (2006). Adsorption of zinc(II) from water with purified carbon nanotubes. *Chemical Engineering Science, 61*(4), 1138–1145.

90. Lu, X. C., Jiang, J. C., Sun, K., Wang, J. B., & Zhang, Y. P., (2014). Influence of the pore structure and surface chemical properties of activated carbon on the adsorption of mercury from aqueous solutions. *Marine Pollution Bulletin, 78*(1, 2), 69–76.

91. Maher, A., Sadeghi, M., & Moheb, A., (2014). Heavy metal elimination from drinking water using nanofiltration membrane technology and process optimization using response surface methodology. *Desalination, 352*, 166–173.

92. Mahmoud, A., & Hoadley, A. F. A., (2012). An evaluation of a hybrid ion exchange electrodialysis process in the recovery of heavy metals from simulated dilute industrial wastewater. *Water Research, 46*(10), 3364–3376.

93. Mata, Y. N., Blazquez, M. L., Ballester, A., Gonzalez, F., & Munoz, J. A., (2010). Studies on sorption, desorption, regeneration and reuse of sugar-beet pectin gels for heavy metal removal. *Journal of Hazardous Materials, 178*(1–3), 243–248.

94. Mata, Y. N., Blazquez, M. L., Ballester, A., Gonzalez, F., & Munoz, J. A., (2009). Sugar-beet pulp pectin gels as biosorbent for heavy metals: Preparation and determination

of biosorption and desorption characteristics. *Chemical Engineering Journal, 150*(2, 3), 289–301.

95. Miretzky, P., & Cirelli, A. F., (2010). Cr(VI) and Cr(III) removal from aqueous solution by raw and modified lignocellulosic materials: A review. *Journal of Hazardous Materials, 180*(1–3), 1–19.

96. Mishra, P. C., & Patel, R. K., (2009). Removal of lead and zinc ions from water by low cost adsorbents. *Journal of Hazardous Materials, 168*(1), 319–325.

97. Mobasherpour, I., Salahi, E., & Ebrahimi, M., (2012). Removal of divalent nickel cations from aqueous solution by multi-walled carbon nano tubes: Equilibrium and kinetic processes. *Research on Chemical Intermediates, 38*(9), 2205–2222.

98. Mohammadi, T., & Kaviani, A., (2003). Water shortage and seawater desalination by electrodialysis. *Desalination, 158*(1–3), 267–270.

99. Mohan, D., Singh, K. P., & Singh, V. K., (2006). Trivalent chromium removal from wastewater using low cost activated carbon derived from agricultural waste material and activated carbon fabric cloth. *Journal of Hazardous Materials, 135*(1–3), 280–295.

100. Moradi, O., (2011). The removal of ions by functionalized carbon nanotube: Equilibrium, isotherms and thermodynamic studies. *Chemical and Biochemical Engineering Quarterly, 25*(2), 229–240.

101. Motsi, T., Rowson, N. A., & Simmons, M. J. H., (2009). Adsorption of heavy metals from acid mine drainage by natural zeolite. *International Journal of Mineral Processing, 92*(1, 2), 42–48.

102. Namasivayam, C., & Kadirvelu, K., (1999). Uptake of mercury (II) from wastewater by activated carbon from an unwanted agricultural solid by-product: Coir pith. *Carbon, 37*(1), 79–84.

103. Nascimento, M., Soares, P. S. M., & De Souza, V. P., (2009). Adsorption of heavy metal cations using coal fly ash modified by hydrothermal method. *Fuel, 88*(9), 1714–1719.

104. Naseem, R., & Tahir, S. S., (2001). Removal of Pb(II) from aqueous/acidic solutions by using bentonite as an adsorbent. *Water Research, 35*(16), 3982–3986.

105. Ngah, W. S. W., & Hanafiah, M. A. K. M., (2008). Removal of heavy metal ions from wastewater by chemically modified plant wastes as adsorbents: A review. *Bioresource Technology, 99*(10), 3935–3948.

106. Ngah, W. S. W., Teong, L. C., & Hanafiah, M. A. K. M., (2011). Adsorption of dyes and heavy metal ions by chitosan composites: A review. *Carbohydrate Polymers, 83*(4), 1446–1456.

107. Nikolaev, P., Bronikowski, M. J., Bradley, R. K., Rohmund, F., Colbert, D. T., Smith, K. A., & Smalley, R. E., (1999). Gas-phase catalytic growth of single-walled carbon nanotubes from carbon monoxide. *Chemical Physics Letters, 313*(1, 2), 91–97.

108. Ntim, S. A., & Mitra, S., (2012). Adsorption of arsenic on multiwall carbon nanotube-zirconia nanohybrid for potential drinking water purification. *Journal of Colloid and Interface Science, 375*, 154–159.

109. Pan, S. C., Lin, C. C., & Tseng, D. H., (2003). Reusing sewage sludge ash as adsorbent for copper removal from wastewater. *Resources Conservation and Recycling, 39*(1), 79–90.

110. Patil, D. S., Chavan, S. M., & Oubagaranadin, J. U. K., (2016). A review of technologies for manganese removal from wastewaters. *Journal of Environmental Chemical Engineering, 4*(1), 468–487.

111. Paulino, A. T., Minasse, F. A. S., Guilherme, M. R., Reis, A. V., Muniz, E. C., & Nozaki, J., (2006). Novel adsorbent based on silkworm chrysalides for removal of heavy metals from wastewaters. *Journal of Colloid and Interface Science, 301*(2), 479–487.

112. Peng, X. J., Luan, Z. K., Ding, J., Di, Z. H., Li, Y. H., & Tian, B. H., (2005). Ceria nanoparticles supported on carbon nanotubes for the removal of arsenate from water. *Materials Letters, 59*(4), 399–403.

113. Peric, J., Trgo, M., & Medvidovic, N. V., (2004). Removal of zinc, copper and lead by natural zeolite - a comparison of adsorption isotherms. *Water Research, 38*(7), 1893–1899.

114. Pillay, K., Cukrowska, E. M., & Coville, N. J., (2009). Multi-walled carbon nanotubes as adsorbents for the removal of parts per billion levels of hexavalent chromium from aqueous solution. *Journal of Hazardous Materials, 166*(2, 3), 1067–1075.

115. Pradhan, D., Sukla, L. B., Sawyer, M., & Rahman, P. K. S. M., (2017). Recent bioreduction of hexavalent chromium in wastewater treatment: A review. *Journal of Industrial and Engineering Chemistry, 55*, 1–20.

116. Qian, D., Wagner, G. J., Liu, W. K., Yu, M. F., & Ruoff, R. S., (2002). Mechanics of carbon nanotubes. *Applied Mechanics Reviews, 55*(6), 495–533.

117. Qu, X. L., Alvarez, P. J. J., & Li, Q. L., (2013). Applications of nanotechnology in water and wastewater treatment. *Water Research, 47*(12), 3931–3946.

118. Rahman, M. S., & Islam, M. R., (2009). Effects of pH on isotherms modeling for Cu(II) ions adsorption using maple wood sawdust. *Chemical Engineering Journal, 149*(1–3), 273–280.

119. Rao, G. P., Lu, C., & Su, F., (2007). Sorption of divalent metal ions from aqueous solution by carbon nanotubes: A review. *Separation and Purification Technology, 58*(1), 224–231.

120. Rao, M., Parwate, A. V., & Bhole, A. G., (2002). Removal of Cr^{6+} and Ni^{2+} from aqueous solution using bagasse and fly ash. *Waste Management, 22*(7), 821–830.

121. Rao, M. M., Ramesh, A., Rao, G. P. C., & Seshaiah, K., (2006). Removal of copper and cadmium from the aqueous solutions by activated carbon derived from *Ceiba pentandra* hulls. *Journal of Hazardous Materials, 129*(1–3), 123–129.

122. Reddad, Z., Gerente, C., Andres, Y., & Le Cloirec, P., (2002). Adsorption of several metal ions onto a low-cost biosorbent: Kinetic and equilibrium studies. *Environmental Science & Technology, 36*(9), 2067–2073.

123. Ren, X. M., Chen, C. L., Nagatsu, M., & Wang, X. K., (2011). Carbon nanotubes as adsorbents in environmental pollution management: A review. *Chemical Engineering Journal, 170*(2, 3), 395–410.

124. Ren, X. M., Li, J. X., Tan, X. L., & Wang, X. K., (2013). Comparative study of graphene oxide, activated carbon and carbon nanotubes as adsorbents for copper decontamination. *Dalton Transactions, 42*(15), 5266–5274.

125. Ren, X. M., Shao, D. D., Zhao, G. X., Sheng, G. D., Hu, J., Yang, S. T., & Wang, X. K., (2011). Plasma induced multiwalled carbon nanotube grafted with 2-vinylpyridine for preconcentration of Pb(II) from aqueous solutions. *Plasma Processes and Polymers, 8*(7), 589–598.

126. Ren, Y., Abbood, H. A., He, F., Peng, H., & Huang, K., (2013). Magnetic EDTA-modified chitosan/SiO_2/Fe_3O_4 adsorbent: Preparation, characterization, and application in heavy metal adsorption. *Chemical Engineering Journal, 226*, 300–311.

127. Ren, Z. F., Huang, Z. P., Xu, J. W., Wang, J. H., Bush, P., Siegal, M. P., & Provencio, P. N., (1998). Synthesis of large arrays of well-aligned carbon nanotubes on glass. *Science, 282*(5391), 1105–1107.

128. Sadrzadeh, M., & Mohammadi, T., (2009). Treatment of sea water using electrodialysis: Current efficiency evaluation. *Desalination, 249*(1), 279–285.

129. Saleh, T., & Gupta, V., (2012). Column with CNT/magnesium oxide composite for lead(II) removal from water. *Environmental Science and Pollution Research, 19*(4), 1224–1228.

130. Sekar, M., Sakthi, V., & Rengaraj, S., (2004). Kinetics and equilibrium adsorption study of lead(II) onto activated carbon prepared from coconut shell. *Journal of Colloid and Interface Science, 279*(2), 307–313.

131. Sellaoui, L., Depci, T., Kul, A. R., Knani, S., & Ben, L. A., (2016). A new statistical physics model to interpret the binary adsorption isotherms of lead and zinc on activated carbon. *Journal of Molecular Liquids, 214*, 220–230.

132. Shadbad, M. J., Mohebbi, A., & Soltani, A., (2011). Mercury(II) removal from aqueous solutions by adsorption on multi-walled carbon nanotubes. *Korean Journal of Chemical Engineering, 28*(4), 1029–1034.

133. ShamsiJazeyi, H., & Kaghazchi, T., (2010). Investigation of nitric acid treatment of activated carbon for enhanced aqueous mercury removal. *Journal of Industrial and Engineering Chemistry, 16*(5), 852–858.

134. Shao, D. D., Hu, J., & Wang, X. K., (2010). Plasma induced grafting multiwalled carbon nanotube with chitosan and its application for removal of $UO2^{2+}$, Cu^{2+}, and Pb^{2+} from aqueous solutions. *Plasma Processes and Polymers, 7*(12), 977–985.

135. Srivastava, N. K., & Majumder, C. B., (2008). Novel biofiltration methods for the treatment of heavy metals from industrial wastewater. *Journal of Hazardous Materials, 151*(1), 1–8.

136. Srivastava, P., Singh, B., & Angove, M., (2005). Competitive adsorption behavior of heavy metals on kaolinite. *Journal of Colloid and Interface Science, 290*(1), 28–38.

137. Tawabini, B., Al-Khaldi, S., Atieh, M., & Khaled, M., (2010). Removal of mercury from water by multi-walled carbon nanotubes. *Water Science and Technology, 61*(3), 591–598.

138. Tawabini, B. S., Al-Khaldi, S. F., Khaled, M. M., & Atieh, M. A., (2011). Removal of arsenic from water by iron oxide nanoparticles impregnated on carbon nanotubes. *Journal of Environmental Science and Health Part A: Toxic/Hazardous Substances & Environmental Engineering, 46*(3), 215–223.

139. Tofighy, M. A., & Mohammadi, T., (2011). Adsorption of divalent heavy metal ions from water using carbon nanotube sheets. *Journal of Hazardous Materials, 185*(1), 140–147.

140. Ulbricht, H., Kriebel, J., Moos, G., & Hertel, T., (2002). Desorption kinetics and interaction of Xe with single-wall carbon nanotube bundles. *Chemical Physics Letters, 363*(3, 4), 252–260.

141. Unlu, N., & Ersoz, M., (2006). Adsorption characteristics of heavy metal ions onto a low cost biopolymeric sorbent from aqueous solutions. *Journal of Hazardous Materials, 136*(2), 272–280.

142. Verma, B., & Balomajumder, C., (2020). Fabrication of magnetic cobalt ferrite nanocomposites: An advanced method of removal of toxic dichromate ions from electroplating wastewater. *Korean Journal of Chemical Engineering, 37*(7), 1157–1165.

143. Verma, B., & Balomajumder, C., (2020). Synthesis of magnetic nickel ferrites nano-composites: An advanced remediation of electroplating wastewater. *Journal of the Taiwan Institute of Chemical Engineers, 112*, 106–115.

144. Verma, B., Sewani, H., & Balomajumder, C., (2020). Synthesis of carbon nano-tubes via chemical vapor deposition: An advanced application in the management of electroplating effluent. *Environmental Science and Pollution Research, 27*(12), 14007–14018.

145. Vukovic, G. D., Marinkovic, A. D., Colic, M., Ristic, M. D., Aleksic, R., Peric-Grujic, A. A., & Uskokovic, P. S., (2010). Removal of cadmium from aqueous solutions by oxidized and ethylenediamine-functionalized multi-walled carbon nanotubes. *Chemical Engineering Journal, 157*(1), 238–248.

146. Vukovic, G. D., Marinkovic, A. D., Skapin, S. D., Ristic, M. D., Aleksic, R., Peric-Grujic, A. A., & Uskokovic, P. S., (2011). Removal of lead from water by amino modified multi-walled carbon nanotubes. *Chemical Engineering Journal, 173*(3), 855–865.

147. Wajima, T., & Sugawara, K., (2011). Adsorption behaviors of mercury from aqueous solution using sulfur-impregnated adsorbent developed from coal. *Fuel Processing Technology, 92*(7), 1322–1327.

148. Wang, H. J., Zhou, A. L., Peng, F., Yu, H., & Chen, L. F., (2007). Adsorption characteristic of acidified carbon nanotubes for heavy metal Pb(II) in aqueous solution. *Materials Science and Engineering a-Structural Materials Properties Microstructure and Processing, 466*(1, 2), 201–206.

149. Wang, Q., Li, J. X., Chen, C. L., Ren, X. M., Hu, J., & Wang, X. K., (2011). Removal of cobalt from aqueous solution by magnetic multiwalled carbon nanotube/iron oxide composites. *Chemical Engineering Journal, 174*(1), 126–133.

150. Wang, S. B., & Peng, Y. L., (2010). Natural zeolites as effective adsorbents in water and wastewater treatment. *Chemical Engineering Journal, 156*(1), 11–24.

151. Wang, S. G., Gong, W. X., Liu, X. W., Yao, Y. W., Gao, B. Y., & Yue, Q. Y., (2007). Removal of lead(II) from aqueous solution by adsorption onto manganese oxide-coated carbon nanotubes. *Separation and Purification Technology, 58*(1), 17–23.

152. Yaacoubi, H., Zidani, O., Mouflih, M., Gourai, M., & Sebti, S., (2014). Removal of cadmium from water using natural phosphate as adsorbent. *Procedia Engineering, 83*, 386–393.

153. Yang, S. T., Li, J. X., Shao, D. D., Hu, J., & Wang, X. K., (2009). Adsorption of Ni(II) on oxidized multi-walled carbon nanotubes: Effect of contact time, pH, foreign ions and PAA. *Journal of Hazardous Materials, 166*(1), 109–116.

154. Yang, S. T., Ren, X. M., Zhao, G. X., Shi, W. Q., Montavon, G., Grambow, B., & Wang, X. K., (2015). Competitive sorption and selective sequence of Cu(II) and Ni(II) on montmorillonite: Batch, modeling, EPR and XAS studies. *Geochimica Et Cosmochimica Acta, 166*, 129–145.

155. Yang, S. T., Sheng, G. D., Tan, X. L., Hu, J., Du, J. Z., Montavon, G., & Wang, X. K., (2011). Determination of Ni(II) uptake mechanisms on mordenite surfaces: A combined macroscopic and microscopic approach. *Geochimica Et Cosmochimica Acta, 75*(21), 6520–6534.

156. Yavuz, O., Altunkaynak, Y., & Guzel, F., (2003). Removal of copper, nickel, cobalt and manganese from aqueous solution by kaolinite. *Water Research, 37*(4), 948–952.

157. Zewail, T. M., & Yousef, N. S., (2015). Kinetic study of heavy metal ions removal by ion exchange in batch conical air spouted bed. *Alexandria Engineering Journal, 54*(1), 83–90.

158. Zhang, C., Sui, J. H., Li, J., Tang, Y. L., & Cai, W., (2012). Efficient removal of heavy metal ions by thiol-functionalized superparamagnetic carbon nanotubes. *Chemical Engineering Journal, 210*, 45–52.

159. Zourmand, Z., Faridirad, F., Kasiri, N., & Mohammadi, T., (2015). Mass transfer modeling of desalination through an electrodialysis cell. *Desalination, 359*, 41–51.

160. Zwain, H. M., Vakili, M., & Dahlan, I., (2014). Waste material adsorbents for zinc removal from wastewater: A comprehensive review. *International Journal of Chemical Engineering, 2014*, 347912.

CHAPTER 12

NANOPARTICLES DERIVED FROM LIGNOCELLULOSIC BIOMASS

AJAY KUMAR CHAUHAN

ABSTRACT

For the nanotechnological world, lignocellulosic biomass provides the sustainable material which is renewable, recyclable, environmentally friendly and cost-effective. Uses of lignin produce the lignin nanoparticles (LNPs), which are used in medical sciences and help in finding the new routes to biorefinery and lignin valorization. Recent research studies on the carbon quantum dots (CQDs) and nanocomposites have shown the potential of the lignocellulosic biomass as the biocompatible material. Various processes of producing nanocellulose and its major limitations of highly energy intensive and for LNPs, easy dispersity and easily accessing of the lignin monomer is the big challenge.

12.1 INTRODUCTION

Nanocellulose is a natural fiber originating from lignocellulosic biomass, having hydroxyl groups that tend to hold strong hydrogen bonding networks with strong physical and mechanical properties. Cellulose fiber has the crystalline and amorphous structure. In the crystalline part, chain molecules are orderly packed to give high strength and stiffness; and amorphous part gives the flexibility of the material [15, 48].

The nanocellulose fiber is easily biodegradable, lighter in weight, density around 1.3 g/ cm^3 making it suitable for nanoparticles (NPs) [41]. These nanocellulose fibers are better alternative to the inorganic nanomaterials to produce functional polymer composites without degradation of the polymer matrix. Other than the cellulose, major focus on lignin and its conversion into NPs is the new challenging area for researchers.

Lignin consists of 15–40% of the dry lignocellulosic biomass. Paper and pulp industry is the main producer of the lignin [40], but it is underutilized for the co-power generation [76]. Lignin is easily available resource on the earth, its process of extraction is the major concern in the environment; it can be recovered in a such a way that its conversion leads to 2% yield of the value-added products [38, 72]; however, its complicated structure and heterogeneity is a biggest challenge [45].

Carbon based nanoparticles (CNP) have shown high electrical and thermal conductivity, high mechanical strength with variable optical properties, dispersibility, higher specific surface area, and flexibility in molecular design [8, 22, 46]. All these properties are used for the preparation of lignin nanoparticles (LNPs) that are nontoxic, low-cost, and environmentally viable and biocompatible [29, 59].

LNPs can be easily modified by the functional groups into the matrix material of lignin and allows the fabrication of LNPs nanocomposite [73]. These LNPs have been used for: delivery of hydrophobic molecule, carbon quantum dots (CQDs) for bioimaging, manufacturing of biosensors (due to optical properties and fluorescence) [67]. LNPs can be precipitated by some anti-solvent [73]. CQDs are prepared by the hydrothermal treatment and microwave irradiation.

Most effective method is microwave irradiation due to fast, feasible, energy intensive, time-saver, and formation of a homogenous product. Si et al. introduced the concept of making LNPs and CQDs for the synchronous preparation of different scaled nanomaterials from lignocellulose [28, 67, 87]. From the structural point of view, LNPs exhibit micelle-shell structure due to the presence of the hydrophobic phenyl-propanoid unit and hydrophilic phenolic and aliphatic parts. Unique micelle-shell nanostructure rich in phenolic hydroxyl groups on the shell allows LNPs to easy dispersion and stability for several months [42].

Enzymatic conversions of lignin into nanoparticles (NPs) requires several enzymes, such as: cellulase, lytic polysaccharide monooxygenase (LPMO), cellobiose dehydrogenase (CDH); and various peroxidases, which lead to several physicochemical changes of enzymes or NPs. These changes (such as: enzyme adsorbed on the surfaces of NPs) cause conformational changes, which either lead inhibition or enhancement of enzymatic activity [16]. Hence various methods for the enzyme immobilization on different supports but immobilizations of the various enzymes are used to convert lignin to intermediate particles of LNPs and cellulose to nanocellulose; and this decreases the enzyme leakage and gives higher bonding with higher product yield [55].

This chapter focuses on: NPs production from lignin via chemical, mechanical, and biological routes; and their feasibility and applications in the nanoworld.

12.2 LIGNOCELLULOSIC BIOMASS

Lignocellulosic biomass is the major portion of plants and non-wood agro-waste material (Table 12.1). Cellulose is composed of β-1-4-glycosidic linkage of the glucose molecule up to 10,000 units of it [39, 61]. These chains give rise to rigid and semi-crystalline structure due to the hydrogen bonds and *Van der Waals* forces interaction; yielding in cellulose fibers that are insoluble in water and conventional solvents [39, 61].

TABLE 12.1 Different Lignocellulosic Biomass and its Constituents

Plant Resource	Cellulose (%)	Hemi-Cellulose (%)	Lignin (%)		References
Miscanthus	45–52	24–33	9–13	Aromatic components	[6]
Switchgrass	32–37	26–33	17–18	Syringic	[11]
Corn stover	37	31	18	Guaiacol	[63]
Poplar	42–48	16–22	21–27	Hydroxyphenyl	[65]
Eucalyptus	39–46	24–28	29–32		[51]
Pine	46	23	28		[31]

Note: Molecular weight for native lignin varies from material to material: 78,400 g/mol in spruce [23] to 8,300 g/mol in miscanthus [17].

Hemicellulose is part of carbohydrates; and the cross-linking glycans comprise of pentosans, hexoses, uronic acids, and acetyl moieties in side-chains. Degree of polymerization (DP) of in the range of 50–300 units make it more susceptible to chemical attack [78, 79]. Total carbohydrates include: cellulose and hemicellulose (named as holocellulase). Lignin is the third constituent and 2nd most source available biomass on earth; and is more complex in structure and recalcitrant in nature.

Lignin mainly comprises of the p-propylphenol polymer that is polymerization of the monolignol forms of lignin [4]. It is a cementing material to provide rigidity to the plant cell-wall and to prevent it from the microbial attack [3, 12]. It is synthesized by the phenylpropanoid biosynthetic pathway. In this process, multi-enzymes are required for the formation of biochemical network of phenyl alanine and tyrosine (grasses), which converts it into the three lignin building blocks, such as: p-coumaryl alcohol, coniferyl alcohol, and sinapyl alcohol. Figure 12.1 shows the extraction of Nanocellulose from lignocellulosic biomass.

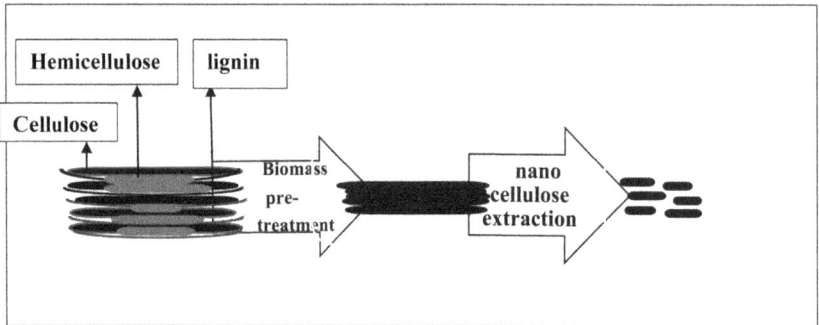

FIGURE 12.1   Nanocellulose extraction from lignocellulosic biomass.

12.3   NANOPARTICLES (NPS) FROM LIGNOCELLULOSIC BIOMASS

12.3.1   NANOCELLULOSE

It is a natural fiber with high stiffness up to 220 GPa and high strength up to 10 GPa greater than the cast iron; and its ratio of strength to weight is 8 times than the stainless steel [55]. Its transparency and presence of hydroxyl groups makes it more reactive. These nanocellulose are classified into four types, such as: nanocrystalline cellulose, nano-fibrillated cellulose, bacterial nanocellulose and hairy cellulose nano-crystalloid. All these are similar in chemical composition. However, these differ in morphology, size, crystallinity, and source of origin and extraction process [41, 48].

12.3.1.1   NANO CRYSTALLINE CELLULOSE

Cellulose nanocrystals or cellulose nano-whiskers are other names of nano crystalline cellulose. The acid treatment of cellulose fibers produces nano crystalline cellulose, after the hydrolysis separation [14, 15]. Its diameter is 2–20 nm and it is 100–500 nm in length with short rod in shape or whisker. It has 100% cellulose with crystallinity of 54–88%.

12.3.1.2   NANO FIBRILLATED CELLULOSE (NFC)

Cellulose microfibril, micro-fibrillated cellulose and cellulose nanofiber or cellulose nanofibrillar are alternate names of the nano-fibrillated fiber. Mechanical treatment causes cleavage in longitudinal direction of the fiber during extraction and it produces the nano fibrillated fiber [48]. Its diameter size is 1–100 nm with length of 500–2,000 nm. It is long, flexible, and entangled nanocellulose [1, 50]. Its chemical composition is 100% cellulose with both crystalline and amorphous regions [14, 15]. Nano-fibrillated cellulose is longer in length and diameter with high surface area compared to nanocrystalline cellulose and it has the most important higher hydroxyl groups with surface modifications [41].

12.3.1.3   BACTERIAL NANOCELLULOSE

This nanocellulose is built up by low molecular weight of sugar utilized by the bacteria (*Glucono acetobacter xylinus*) [1, 35, 50]. This type of nanocellulose is free from impurities and it is found in twisting ribbons with average diameter of 2–100 nm and higher strength properties [35].

12.3.1.4   HAIRY CELLULOSE NANO-CRYSTALLOID

Hairy cellulose nano-crystalloid (HCNC) is extracted from the cellulose chains by chemical treatment method (acid hydrolysis and mechanical methods). Chemical treatment changes the amorphous region by removing and solubilization; and crystalline regions remain like extraction of the nanocrystalline cellulose. The form of these HCNC are cellulose chains protruded and cleaved from both ends, so that HCNC has both amorphous and crystalline parts. On the basis of the protruded tails, its derivative is subclassified as sterically stabilized nanocrystalline cellulose (SNCC) and electro-sterically stabilized nanocrystalline cellulose (ENCC).

The SNCC can be produced by the reaction of the periodate (IO_4^-) followed by heat treatment, thus leading to negative charges on the periodate ions; and it has the potential of breaking C_2–C_3 bonds of the glucose unit. It also converts C_2–C_3 hydroxyl groups into aldehydes, when di-aldehyde chains are cleaved and created at the end of crystalline regions, thus resulting in better penetration at end of the amorphous region of cellulosic chain [55, 77, 83].

The electro-sterically stabilized nanocrystalline cellulose (ENCC) is extracted from the cellulosic chains by the oxidative action of the periodate or chlorite. Rod shaped crystalline parts are further extracted and protruded by decarboxylated chains, which along with ENCC have highly charged and higher colloidal stability than the other nanocrystalline cellulose [62, 71, 77, 83].

12.3.2   ENZYMATIC FRACTIONATION OF RECALCITRANT CELLULOSE

This is eco-friendly and alternate process compared to the chemically synthesized nanocellulose, which is combined with the mechanical homogenization [26, 27, 53]. The hydrolytic enzymes (such as: endoglucanase [27, 53], exoglucanase [26] and β-glucosidase) are used for the enzymatic fractionation of the recalcitrant cellulose.

By using cellulase, significant reduction in DP from 1,400 to 400 in 48 h of hydrolysis followed by mechanical homogenization leads to formation of nano-fibrillate cellulose, which when assisted with pre-treatment of recalcitrant cellulose has shown proven potency in cutting of cellulose chains and fibers [89]. Other enzymes showing the reduction in cellulose DP are: LPMO [9] cleaved at the C_1 and C_4 (or both); CDH; and these can shorten the fiber length for producing nanocellulose fiber [70].

12.3.3   LIGNIN NANOPARTICLES (LNPS)

Production of the LNPs is the alternate approach for lignin valorization [49, 58]. Similar to other bioresources (such as: cellulose, chitosan, and starch, etc. [7, 24]), LNPs are natural polymeric nanomaterials and show potential for fabricating engineered nanocomposites [59, 88]. Fabricating LNPs is easier compared to lignin valorization that is very controllable process yielding uniform formation of product from the complex and heterogenous lignin materials (Figure 12.2).

In the present strategy of LNPs production by solution based micelliza-tion, lignin is first dissolved in the selected organic solvent, and dissolved lignin is used to precipitate by the dropwise addition or dialysis by using antisolvent [56, 74]. Further micellization process is used for the phase separation due to the amphiphilic property of lignin. This process is similar to make amphiphilic di-block co-polymer, where hydrophobic part is made up phenyl propanoid unit forming aggregation in the form of micelle core; while hydrophilic part in is made of hydroxyl and carboxyl groups forming micelle shells [86].

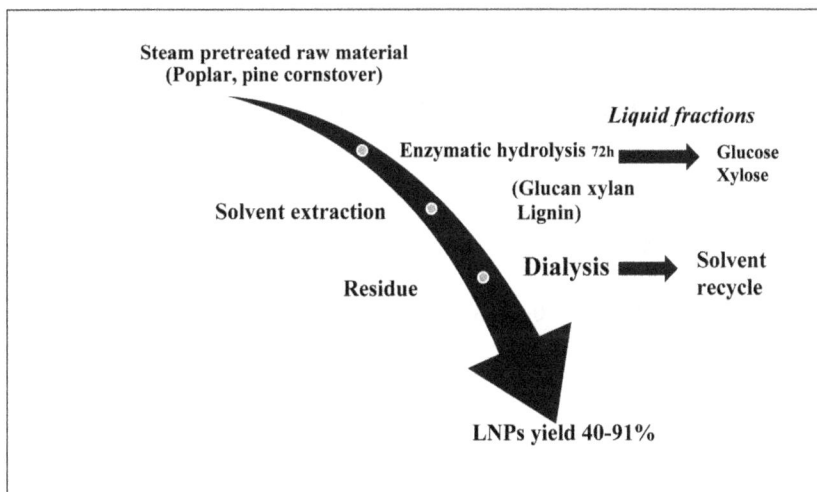

FIGURE 12.2   Flowchart of lignin nanoparticle production.

Tian et al. [73, 74] suggested that if acetylation step is used before micellization, then it increases the hydrophobicity of the lignin material for producing high quality lignin colloidal NPs. When lignocellulosic biomass is used for the bioethanol production, then lignin residue is left after the enzymatic hydrolysis (Cellulase), and steam pre-treatment of lignin opens up the biomass structure to increase the substrate accessibility towards enzymes; and this process causes lignin condensation that is the barrier for the lignin valorization.

12.4   METHODS OF PREPARATION OF NANOCELLULOSE

12.4.1   EXTRACTION OF NANOCELLULOSE BY ACID HYDROLYSIS

Acid hydrolysis requires 47% of sulfuric acid for treating the cellulosic biomass. After treatment acid is removed by centrifugation and is washed with the deionized water, then the product stream is neutralized with 0.5 N sodium hydroxide followed by washing again with the distilled water [47]. Sulfuric acid has affinity towards the amorphous structure of the cellulosic chains; and it can easily hydrolyze the cellulosic chains thus leaving crystalline part of the cellulosic chains [41, 48]. During hydrolysis of cellulosic structure, formation of nanocrystalline cellulose occurs to form nanocellulose dispersed colloidal solution [13, 21] due to the esterification because of sulfate ions. This process has main drawback of generation of acid waste water.

12.4.2   TEMPO-MEDIATED OXIDATION OF CELLULOSE

Oxidants (such as: 2,2,6,6-tetramethylpiperidine-1-oxyl radical known as tempo) are used as the catalysts in the conversion of cellulose into nanocellulose [32]. It acts as the primary oxidant when hypochlorite is used for the oxidation of the hydroxyl group of cellulose into carboxylates [20, 32]. This forms the nano-fibrillated cellulose. Tempo mediated nanofibers have uniform width of fibers of 3–4 nm with transparent property making it "flexible display" that is used in the packaging for gas carrier films and nanofiber filling materials [20, 32].

12.4.3   MICROWAVE ASSISTED PRE-TREATMENT WITH ETHANOL/WATER CO-SOLVENT SYSTEM

It provides rapid treatment of lignocellulosic biomass in a short period; and further ethanol-water co-solvent mixture is supplied with an acid catalyst.

It generates uniform, opaque colloidal solution, which is observed by low-magnification transmission electron microscopy (TEM). Clear visualization shows two types NPs with diameters of 100 nm and 2–3 nm of nanodots. These NPs can be easily separated by centrifugation [67].

12.4.4 ENZYMATIC HYDROLYSIS

In the biological process, enzyme is required to digest the cellulose fibers. This process takes long time for operation [37]. Therefore, the enzymatic process is modified with the incorporation of other method. First lignocellulosic biomass is pre-treated with ionic liquid to enhance the accessible surface area of cellulose; then the process is followed by the enzyme treatment, such as, cellulase, laccase, LPMO or CDH. After this enzymatic treatment, nanocellulose is obtained showing greater crystallinity and thermal stability than the natural fibers obtained from the wood [55].

12.4.5 MECHANICAL TREATMENT

High shear force is required for the cleavage of cellulose fiber in the longitudinal direction, which produces cellulose nanofibers. Mechanical treatment involves high pressure homogenization (HPH), ultrasonication, and ball milling method. In HPH, cellulose slurry is passed into vessel at high pressure and velocity. This generates high pressure shear and causes generation of cellulosic microfibrils of nanosizes [37]. The drawback of the process is that it generates nanocellulose of low crystallinity with low thermal stability.

Ultrasonication defibrillation requires hydrodynamic forces of the ultrasound [37]. In this process, mechanical oscillations result in the formation, expansion, and the generation of the microscopic bubbles due to liquid molecules absorbing ultrasonic energy. This process is operated for 5 h and 70°C to generate the nano-cellulose with yield of ~85% and diameter of 10–100 nm.

In ball milling mechanical process, the centrifugal force of the rotating jar produces the shear forces to crack the cellulose fiber due to the forces generated between the balls, balls, and surface of the jar [2, 18]. This results in small size of nanocellulose. However, this mechanical process is highly energy intensive process, which can be modified with other processes to reduce the energy consumption.

12.4.6 NANOPARTICLES (NPS) PREPARATION FROM ALKALINE LIGNIN (AL)

Alkaline lignin (AL) is dissolved in the organic solvent mixture, which contains methanol, ethanol, and tetrahydrofuran @ 6 mg/mL w/v at room temperature. The insoluble part is separated from the mixture and solution mixture is observed with UV-vis spectrometer. Then mixture is kept on the magnetic stirrer and is diluted. Further, dropwise deionized water is added to continuous vertexing mixture till the concentration of water has reached to 50–90%. This forms the AL NPs. Similarly, resveratrol (rsv) and AL are dissolved in organic solvent and deionized water is replaced with Fe_3O_4 that is used for the preparation of AL-rsv-Fe_3O_4 NPs [10].

12.4.7 PREPARATION OF LIGNIN NANOCOMPOSITES

Lignin is isolated from lignocellulosic biomass and is dissolved in the suitable solvent, such as: di-methyl-sulfoxide (DMSO) @ 2 mg/mL w/v. This lignin forms the micellization by using dialysis to yield uniform LNPs, which was further confirmed by using atomic force microscopy (AFM). LNPs thus obtained have ultraviolet (UV) shielding and antioxidant properties blended with polyvinyl alcohol thus forming nanocomposites with enhanced properties [73].

12.5 APPLICATIONS OF LIGNIN-BASED NANOPARTICLES (LNPS)

NPs from lignocellulosic biomass are biodegradable in nature thus without affecting cost of the environmental cleaning process. LNPs have high mechanical strength and thermal properties with light weight; and it is used in manufacturing of nanocomposite materials, surface modified special materials and transparent papers [1]. Nanocellulose is used as filler in the polymer matrix due to the strength properties. Light weight and high strength properties are used to build wind mill blade, light blade armor and the parts of batteries, etc. [5, 69, 81].

By adding nanocellulose fiber extracted from soybean with synthetic fiber, strength can be significantly increased. Polylactic acid (PLA) along with the nanocellulose is used for the new reinforced nanocomposite material with enhanced interaction with the matrix [60]. Surface modification of the nanocellulose can be achieved by chemical treatment or covalently attached hydroxyl groups, and polymer grafting onto the nanocellulose [14, 48].

Surface modification of nanocellulose is used prominently for the fabrication of the surface, which helps in the protection from polar and non-polar liquids [71]; and this anti-wetting property makes nanocellulose with special properties of self-cleaning, antimicrobial, antireflective, and anticorrosive [44, 68, 71].

Fabrication of nanocellulose-modified filter paper was reported by Phanthong et al. [54]; and this paper was coated with nanocellulose and further chemical vapor deposition (CVD) was done with trichloro silane (1H,1H,2H,2H-tridecafluoro-n-octyl). They observed that obtained nanocellulose coated filter paper had higher hydrophobicity and oleopho-bicity repellent from the polar and non-polar liquids in a wide range of environment. Other properties, such as, transparency, mechanical strength and stiffness makes it suitable to be applied on the electronic gadgets, solar-cells, and panels, and integrated circuits [30, 33, 52, 64]. Its low toxicity, biocompatibility allows nanocellulose to be used in the field of medical sciences [1]; for example: nano-fibrillated cellulose is used for the dressing of the wounds [1, 25]. The biocompatibility property of nanocel-lulose is used for the drug delivery system, tissue implants, blood vessel replacements [34, 43].

Hairy cellulose nano-crystalloids are used for the nanocarrier of medi-cine in the targeted drug delivery system. Electro-sterically stabilized nanocrystalline cellulose (ENCC) is used for packaging of food material [85]. ENCC with highly decarboxylated surface modification is used for the removal of heavy metals from waste water treatment [66] (e.g., removal of copper ions @ 185 mg/g has been reported). Crosslinking of the HNCC and carboxymethylated chitosan is used to manufacture biodegradable aerogels, which are highly porous in nature and negatively charged [84].

Other kind of NPs produced from LNPs is a combination with PVA (0–4% w/w of PVA) showing strong shielding against the UV light with antioxidant performance [80, 82]. Some NPs formed from zinc and titanium dioxide cause polymer degradation due to photocatalytic activity, which can be suitably replaced by using LNPs to produce functional protective nanocomposites [36, 57, 75, 80]. Lack of toxicity and easy availability makes lignin as a potential carrier in drug delivery system [19].

Lignin based on the magnetic NPs was successfully loaded with insoluble anticancer drug Rsv by Lin-Dai et al. [10]. LNPs having zero dimension (~CQDs) from the carbon family are used for the bioimaging, drug delivery, and biosensors due to their optical properties that are produced from dissolved lignin in suitable solvent with the microwave irradiation [67].

12.6 SUMMARY

This chapter focuses on how NPs are formed from the lignocellulosic biomass components. NPs preparations by various methods from these components give excellent properties and have its applications in various fields. Such LNPs are less costly, sustainable, having enhanced properties with no toxic effects. Recalcitrant nature of lignin requires high skills and better understanding of enzymatic and chemical catalytic to be converted into the NPs. Separation of the crystalline part from the amorphous part of cellulose is the key challenge for the production of nanocellulose.

KEYWORDS

- carbon quantum dots
- lignin
- lignocellulose
- lytic polysaccharide monooxygenase
- nanocellulose
- nanofibers
- nanoparticles

REFERENCES

1. Abitbol, T., Rivkin, A., Cao, Y., Nevo, Y., Abraham, E., Ben-Shalom, T., Lapidot, S., & Shoseyov, O., (2016). Nanocellulose, a tiny fiber with huge applications. *Current Opinion in Biotechnology, 39*, 76–88.
2. Barakat, A., Mayer-Laigle, C., Solhy, A., Arancon, R. A., De Vries, H., & Luque, R., (2014). Mechanical pretreatments of lignocellulosic biomass: Towards facile and

environmentally sound technologies for biofuels production. *RSC Advances, 4*(89), 48109–48127.

3. Barros, J., Serrani-Yarce, J. C., Chen, F., Baxter, D., Venables, B. J., & Dixon, R. A., (2016). Role of bifunctional ammonia-lyase in grass cell wall biosynthesis. *Nature Plants, 2*(6), 8. Article ID: 16050.

4. Boerjan, W., Ralph, J., & Baucher, M., (2003). Lignin biosynthesis. *Annual Review of Plant Biology, 54*(1), 519–546.

5. Bras, J., Hassan, M. L., Bruzesse, C., Hassan, E. A., El-Wakil, N. A., & Dufresne, A., (2010). Mechanical, barrier, and biodegradability properties of bagasse cellulose whiskers reinforced natural rubber nanocomposites. *Industrial Crops and Products, 32*(3), 627–633.

6. Brosse, N., Dufour, A., Meng, X., Sun, Q., & Ragauskas, A., (2012). Miscanthus: A fast-growing crop for biofuels and chemicals production. *Biofuels, Bioproducts and Biorefining, 6*(5), 580–598.

7. Cao, J., Zhang, X., Wu, X., Wang, S., & Lu, C., (2016). Cellulose nanocrystals mediated assembly of graphene in rubber composites for chemical sensing applications. *Carbohydrate Polymers, 140*, 88–95.

8. Cazacu, G., Pascu, M. C., Profire, L., Kowarski, A. I., Mihaes, M., & Vasile, C., (2004). Lignin role in a complex polyolefin blend. *Industrial Crops and Products, 20*(2), 261–273.

9. Chabbert, B., Habrant, A., Herbaut, M., Foulon, L., Aguié-Béghin, V., Garajova, S., Grisel, S., et al., (2017). Action of lytic polysaccharide monooxygenase on plant tissue is governed by cellular type. *Scientific Reports, 7*(1), 7. Article ID: 17792.

10. Dai, L., Liu, R., Hu, L. Q., Zou, Z. F., & Si, C. L., (2017). Lignin nanoparticle as a novel green carrier for the efficient delivery of resveratrol. *ACS Sustainable Chemistry & Engineering, 5*(9), 8241–8249.

11. David, K., & Ragauskas, A. J., (2010). Switch grass as an energy crop for biofuel production: A review of its ligno-cellulosic chemical properties. *Energy & Environmental Science, 3*(9), 1182–1190.

12. Davison, B. H., Parks, J., Davis, M. F., & Donohoe, B. S., (2013). Plant cell walls: Basics of structure, chemistry, accessibility, and the influence on conversion. *Aqueous Pretreatment of Plant Biomass for Biological and Chemical Conversion to Fuels and Chemicals, 2013*, 23–38.

13. Dong, X. M., Revol, J. F., & Gray, D. G., (1998). Effect of microcrystallite preparation conditions on the formation of colloid crystals of cellulose. *Cellulose, 5*(1), 19–32.

14. Dufresne, A., (2013). Nanocellulose: A new ageless bionanomaterial. *Materials Today, 16*(6), 220–227.

15. Dufresne, A., (2012). Nanocellulose: Potential reinforcement in composites. *Natural Polymers: Nanocomposites, 2*, 1–32.

16. Dutta, N., & Saha, M. K., (2018). Nanoparticle induced enzyme pre-treatment method for increased glucose production from lignocellulosic biomass under cold conditions. *Journal of the Science of Food and Agriculture, 99*(2), 9. Article ID: 9245. doi: 10.1002/jsfa.9245.

17. El Hage, R., Brosse, N., Chrusciel, L., Sanchez, C., Sannigrahi, P., & Ragauskas, A., (2009). Characterization of milled wood lignin and ethanol organosolv lignin from *Miscanthus*. *Polymer Degradation and Stability, 94*(10), 1632–1638.

18. Feng, Y., Han, K., & Owen, D., (2004). Discrete element simulation of the dynamics of high energy planetary ball milling processes. *Materials Science and Engineering: A, 375*, 815–819.

19. Figueiredo, P., Lintinen, K., Kiriazis, A., Hynninen, V., Liu, Z., Bauleth-Ramos, T., Rahikkala, A., et al., (2017). *In vitro* evaluation of biodegradable lignin-based nanoparticles for drug delivery and enhanced antiproliferation effect in cancer cells. *Biomaterials, 121*, 97–108.

20. Fukuzumi, H., Saito, T., Iwata, T., Kumamoto, Y., & Isogai, A., (2008). Transparent and high gas barrier films of cellulose nanofibers prepared by TEMPO-mediated oxidation. *Biomacromolecules, 10*(1), 162–165.

21. George, J., & Sabapathi, S., (2015). Cellulose nanocrystals: Synthesis, functional properties, and applications. *Nanotechnology, Science and Applications, 8*, 45–52.

22. Gosselink, R. J. A., Snijder, M. H. B., Kranenbarg, A., Keijsers, E. R. P., De Jong, E., & Stigsson, L. L., (2004). Characterization and application of novafiber lignin. *Industrial Crops and Products, 20*(2), 191–203.

23. Guerra, A., Filpponen, I., Lucia, L. A., Saquing, C., Baumberger, S., & Argyropoulos, D. S., (2006). Toward a better understanding of the lignin isolation process from wood. *Journal of Agricultural and Food Chemistry, 54*(16), 5939–5947.

24. Habibi, Y., Lucia, L. A., & Rojas, O. J., (2010). Cellulose nanocrystals: Chemistry, self-assembly, and applications. *Chemical Reviews, 110*(6), 3479–3500.

25. Hakkarainen, T., Koivuniemi, R., Kosonen, M., Escobedo-Lucea, C., Sanz-Garcia, A., Vuola, J., Valtonen, J., et al., (2016). Nanofibrillar Cellulose wound dressing in skin graft donor site treatment. *Journal of Controlled Release, 244*, 292–301.

26. Hayashi, N., Kondo, T., & Ishihara, M., (2005). Enzymatically produced nano-ordered short elements containing cellulose Iβ crystalline domains. *Carbohydrate Polymers, 61*(2), 191–197.

27. Henriksson, M., Henriksson, G., Berglund, L., & Lindström, T., (2007). An environmentally friendly method for enzyme-assisted preparation of microfibrillated cellulose (MFC) nanofibers. *European Polymer Journal, 43*(8), 3434–3441.

28. Hou, J., Yan, J., Zhao, Q., Li, Y., Ding, H., & Ding, L., (2013). Novel one-pot route for large-scale preparation of highly photoluminescent carbon quantum dots powders. *Nanoscale, 5*(20), 9558–9561.

29. Hu, B., Wang, K., Wu, L., Yu, S. H., Antonietti, M., & Titirici, M. M., (2010). Engineering carbon materials from the hydrothermal carbonization process of biomass. *Advanced Materials, 22*(7), 813–828.

30. Hu, L., Zheng, G., Yao, J., Liu, N., Weil, B., Eskilsson, M., Karabulut, E., Ruan, Z., Fan, S., & Bloking, J. T., (2013). Transparent and conductive paper from nanocellulose fibers. *Energy & Environmental Science, 6*(2), 513–518.

31. Huang, F., & Ragauskas, A., (2013). Extraction of hemicellulose from loblolly pine woodchips and subsequent kraft pulping. *Industrial & Engineering Chemistry Research, 52*(4), 1743–1749.

32. Isogai, A., Saito, T., & Fukuzumi, H., (2011). TEMPO-oxidized cellulose nanofibers. *Nanoscale, 3*(1), 71–85.

33. Iwamoto, S., Nakagaito, A. N., Yano, H., & Nogi, M., (2005). Optically transparent composites reinforced with plant fiber-based nanofibers. *Applied Physics A, 81*(6), 1109–1112.

34. Jorfi, M., & Foster, E. J., (2015). Recent advances in Nanocellulose for biomedical applications. *Journal of Applied Polymer Science, 132*(14), 78–88.
35. Jozala, A. F., De Lencastre-Novaes, L. C., Lopes, A. M., De Santos-Ebinuma, V. C., Mazzola, P. G., Pessoa-Jr, A., Grotto, D., et al., (2016). Bacterial nanocellulose production and application: A 10-year overview. *Applied Microbiology and Biotechnology, 100*(5), 2063–2072.
36. Kai, D., Tan, M. J., Chee, P. L., Chua, Y. K., Yap, Y. L., & Loh, X. J., (2016). Towards lignin-based functional materials in a sustainable world. *Green Chemistry, 18*(5), 1175–1200.
37. Khalil, H. A., Davoudpour, Y., Islam, M. N., Mustapha, A., Sudesh, K., Dungani, R., & Jawaid, M., (2014). Production and modification of nanofibrillated cellulose using various mechanical processes: A review. *Carbohydrate Polymers, 99,* 649–665.
38. Kleinert, M., & Barth, T., (2008). Towards a lignincellulosic biorefinery: Direct one-step conversion of lignin to hydrogen-enriched biofuel. *Energy & Fuels, 22*(2), 1371–1379.
39. Klemm, D., Heublein, B., Fink, H. P., & Bohn, A., (2005). Cellulose: Fascinating biopolymer and sustainable raw material. *Angewandte Chemie International Edition, 44*(22), 3358–3393.
40. Laurichesse, S., & Avérous, L., (2014). Chemical modification of lignins: Towards biobased polymers. *Progress in Polymer Science, 39*(7), 1266–1290.
41. Lavoine, N., Desloges, I., Dufresne, A., & Bras, J., (2012). Microfibrillated cellulose-its barrier properties and applications in cellulosic materials: A review. *Carbohydrate Polymers, 90*(2), 735–764.
42. Lievonen, M., Valle-Delgado, J. J., Mattinen, M. L., Hult, E. L., Lintinen, K., Kostiainen, M. A., Paananen, A., et al., (2016). A simple process for lignin nanoparticle preparation. *Green Chemistry, 18*(5), 1416–1422.
43. Lin, N., & Dufresne, A., (2014). Nanocellulose in biomedicine: Current status and future prospect. *European Polymer Journal, 59,* 302–325.
44. Liu, K., Tian, Y., & Jiang, L., (2013). Bio-inspired superoleophobic and smart materials: Design, fabrication, and application. *Progress in Materials Science, 58*(4), 503–564.
45. Liu, W. J., Jiang, H., & Yu, H. Q., (2015). Thermochemical conversion of lignin to functional materials: A review and future directions. *Green Chemistry, 17*(11), 4888–4907.
46. Ma, Q., Yu, Y., Sindoro, M., Fane, A. G., Wang, R., & Zhang, H., (2017). Carbon-based functional materials derived from waste for water remediation and energy storage. *Advanced Materials, 29*(13), 9. Article ID: 1605361.
47. Maiti, S., Jayaramudu, J., Das, K., Reddy, S. M., Sadiku, R., Ray, S. S., & Liu, D., (2013). Preparation and characterization of nano-cellulose with new shape from different precursor. *Carbohydrate Polymers, 98*(1), 562–567.
48. Moon, R. J., Martini, A., Nairn, J., Simonsen, J., & Youngblood, J., (2011). Cellulose nanomaterials review: Structure, properties and nanocomposites. *Chemical Society Reviews, 40*(7), 3941–3994.
49. Myint, A. A., Lee, H. W., Seo, B., Son, W. S., Yoon, J., Yoon, T. J., Park, H. J., et al., (2016). One pot synthesis of eco-friendly lignin nanoparticles with compressed liquid carbon dioxide as an antisolvent. *Green Chemistry, 18*(7), 2129–2146.
50. Nechyporchuk, O., Belgacem, M. N., & Bras, J., (2016). Production of cellulose nanofibrils: A review of recent advances. *Industrial Crops and Products, 93,* 2–25.

51. Nunes, C. A., Lima, C. F., Barbosa, L., Colodette, J. L., & Fidêncio, P. H., (2011). *Determinação de Constituintes Químicos em Madeira de Eucalipto por Pi-CG/EM e Calibração Multivariada: Comparação entre Redes Neurais Artificiais e Máquinas de Vetor Suporte* (Determination of chemical constituents in eucalyptus wood by Pi-CG / EM and multivariate calibration: Comparison between artificial neural networks and support vector machines). *Quim. Nova, 34*(2), 279–283.

52. Okahisa, Y., Yoshida, A., Miyaguchi, S., & Yano, H., (2009). Optically transparent wood-cellulose nanocomposite as a base substrate for flexible organic light-emitting diode displays. *Composites Science and Technology, 69*(11, 12), 1958–1961.

53. Pääkkö, M., Ankerfors, M., Kosonen, H., Nykänen, A., Ahola, S., Österberg, M., Ruokolainen, J., et al., (2007). Enzymatic hydrolysis combined with mechanical shearing and high-pressure homogenization for nanoscale cellulose fibrils and strong gels. *Biomacromolecules, 8*(6), 1934–1941.

54. Phanthong, P., Guan, G., Karnjanakom, S., Hao, X., Wang, Z., Kusakabe, K., & Abudula, A., (2016). Amphiphilic nanocellulose-modified paper: Fabrication and evaluation. *RSC Advances, 6*(16), 13328–13334.

55. Phanthong, P., Reubroycharoen, P., Hao, X., Xu, G., Abudula, A., & Guan, G., (2018). Nanocellulose: Extraction and application. *Carbon Resources Conversion, 1*(1), 32–43.

56. Qian, Y., Deng, Y., Qiu, X., Li, H., & Yang, D., (2014). Formation of uniform colloidal spheres from lignin, a renewable resource recovered from pulping spent liquor. *Green Chemistry, 16*(4), 2156–2163.

57. Ren, J., Wang, S., Gao, C., Chen, X., Li, W., & Peng, F., (2015). TiO_2-containing PVA/ xylan composite films with enhanced mechanical properties, high hydrophobicity and UV shielding performance. *Cellulose, 22*(1), 593–602.

58. Richter, A. P., Bharti, B., Armstrong, H. B., Brown, J. S., Plemmons, D., Paunov, V. N., Stoyanov, S. D., & Velev, O. D., (2016). Synthesis and characterization of biodegradable lignin nanoparticles with tunable surface properties. *Langmuir, 32*(25), 6468–6477.

59. Richter, A. P., Brown, J. S., Bharti, B., Wang, A., Gangwal, S., Houck, K., Hubal, E. A. C., et al., (2015). Environmentally benign antimicrobial nanoparticle based on a silver-infused lignin core. *Nature Nanotechnology, 10*(9), 817–830.

60. Robles, E., Urruzola, I., Labidi, J., & Serrano, L., (2015). Surface-modified nano-cellulose as reinforcement in poly (lactic acid) to conform new composites. *Industrial Crops and Products, 71*, 44–53.

61. Rojas, O. J., (2016). *Cellulose Chemistry and Properties: Fibers, Nanocelluloses and Advanced Materials* (Vol. 271, p. 341). Cham: Springer.

62. Safari, S., Sheikhi, A., & Van De, V. T. G., (2014). Electroacoustic characterization of conventional and electrosterically stabilized nanocrystalline celluloses. *Journal of Colloid and Interface Science, 432*, 151–157.

63. Saha, B. C., Yoshida, T., Cotta, M. A., & Sonomoto, K., (2013). Hydrothermal pretreatment and enzymatic saccharification oof corn stover for efficient ethanol production. *Industrial Crops and Products, 44*, 367–372.

64. Salas, C., Nypelö, T., Rodriguez-Abreu, C., Carrillo, C., & Rojas, O. J., (2014). Nanocellulose properties and applications in colloids and interfaces. *Current Opinion in Colloid & Interface Science, 19*(5), 383–396.

65. Sannigrahi, P., Ragauskas, A. J., & Tuskan, G. A., (2010). Poplar as a feedstock for biofuels: A review of compositional characteristics. *Biofuels, Bioproducts and Biorefining, 4*(2), 209–226.

66. Sheikhi, A., Safari, S., Yang, H., & Van De, V. T. G., (2015). Copper removal using electrosterically stabilized nanocrystalline cellulose. *ACS Applied Materials & Interfaces, 7*(21), 11301–11308.

67. Si, M., Zhang, J., He, Y., Yang, Z., Yan, X., Liu, M., Zhuo, S., et al., (2018). Synchronous and rapid preparation of lignin nanoparticles and carbon quantum dots from natural lignocellulose. *Green Chemistry, 20*(15), 9. doi: 10.1039/C8GC00744F.

68. Si, Y., & Guo, Z., (2015). Superhydrophobic nanocoatings: From materials to fabrications and to applications. *Nanoscale, 7*(14), 5922–5946.

69. Siqueira, G., Bras, J., & Dufresne, A., (2010). Cellulosic bionanocomposites: A review of preparation, properties and applications. *Polymers, 2*(4), 728–765.

70. Tan, T. C., Kracher, D., Gandini, R., Sygmund, C., Kittl, R., Haltrich, D., Hällberg, B. M., et al., (2015). Structural basis for cellobiose dehydrogenase action during oxidative cellulose degradation. *Nature Communications, 6,* 7. Article ID: 7542.

71. Tejado, A., Alam, M. N., Antal, M., Yang, H., & Van De, V. T. G., (2012). Energy requirements for the disintegration of cellulose fibers into cellulose nanofibers. *Cellulose, 19*(3), 831–842.

72. Thakur, V. K., Thakur, M. K., Raghavan, P., & Kessler, M. R., (2014). Progress in green polymer composites from lignin for multifunctional applications: A review. *ACS Sustainable Chemistry & Engineering, 2*(5), 1072–1092.

73. Tian, D., Hu, J., Bao, J., Chandra, R. P., Saddler, J. N., & Lu, C., (2017). Lignin valorization: Lignin nanoparticles as high-value bio-additive for multifunctional nanocomposites. *Biotechnology for Biofuels, 10*(1), 192–199.

74. Tian, D., Hu, J., Chandra, R. P., Saddler, J. N., & Lu, C., (2017). Valorizing recalcitrant cellulolytic enzyme lignin via lignin nanoparticles fabrication in an integrated biorefinery. *ACS Sustainable Chemistry & Engineering, 5*(3), 2702–2710.

75. Tu, Y., Zhou, L., Jin, Y. Z., Gao, C., Ye, Z. Z., Yang, Y. F., & Wang, Q. L., (2010). Transparent and flexible thin films of ZnO-polystyrene nanocomposite for UV-shielding applications. *Journal of Materials Chemistry, 20*(8), 1594–1599.

76. Tuck, C. O., Pérez, E., Horváth, I. T., Sheldon, R. A., & Poliakoff, M., (2012). Valorization of biomass: Deriving more value from waste. *Science, 337*(6095), 695–699.

77. Van De, V. T. G., & Sheikhi, A., (2016). Hairy cellulose nanocrystalloids: A novel class of nanocellulose. *Nanoscale, 8*(33), 15101–15114.

78. Vanzin, G. F., Madson, M., Carpita, N. C., Raikhel, N. V., Keegstra, K., & Reiter, W. D., (2002). The mur2 mutant of *Arabidopsis thaliana* lacks fucosylated xyloglucan because of a lesion in fucosyltransferase AtFUT1. *Proceedings of the National Academy of Sciences, 99*(5), 3340–3345.

79. Vergara, C. E., & Carpita, N. C., (2001). β-D-glycan synthases and the CesA gene family: Lessons to be learned from the mixed-linkage $(1 \rightarrow 3),(1 \rightarrow 4)$ β-D-glucan synthase. In: *Plant Cell Walls* (pp. 145–160). Cham: Springer.

80. Wang, Y., Li, T., Ma, P., Bai, H., Xie, Y., Chen, M., & Dong, W., (2016). Simultaneous enhancements of UV-shielding properties and photostability of Poly (vinyl alcohol) via Incorporation of Sepia eumelanin. *ACS Sustainable Chemistry & Engineering, 4*(4), 2252–2258.

81. Wei, H., Rodriguez, K., Renneckar, S., & Vikesland, P. J., (2014). Environmental science and engineering applications of nanocellulose-based nanocomposites. *Environmental Science: Nano, 1*(4), 302–316.

82. Xiong, S., Wang, Y., Yu, J., Chen, L., Zhu, J., & Hu, Z., (2014). Polydopamine particles for next-generation multifunctional biocomposites. *Journal of Materials Chemistry A, 2*(20), 7578–7587.

83. Yang, H., Chen, D., & Van De, V. T. G., (2015). Preparation and characterization of sterically stabilized nanocrystalline cellulose obtained by periodate oxidation of cellulose fibers. *Cellulose, 22*(3), 1743–1752.

84. Yang, H., Sheikhi, A., & Van De, V. T. G., (2016). Reusable green aerogels from cross-linked hairy nanocrystalline cellulose and modified chitosan for dye removal. *Langmuir, 32*(45), 11771–11779.

85. Yang, H., Tejado, A., Alam, N., Antal, M., & Van De, V. T. G., (2012). Films prepared from electrosterically stabilized nanocrystalline cellulose. *Langmuir, 28*(20), 7834–7842.

86. Zhang, L., Shen, H., & Eisenberg, A., (1997). Phase separation behavior and crew-cut micelle formation of polystyrene-b-poly (acrylic acid) copolymers in solutions. *Macromolecules, 30*(4), 1001–1011.

87. Zhang, X., Jiang, M., Niu, N., Chen, Z., Li, S., Liu, S., & Li, J., (2018). Natural-product-derived carbon dots: From natural products to functional materials. *ChemSusChem, 11*(1), 11–24.

88. Zhao, W., Simmons, B., Singh, S., Ragauskas, A., & Cheng, G., (2016). From lignin association to nano-/micro-particle preparation: Extracting higher value of lignin. *Green Chemistry, 18*(21), 5693–5700.

89. Zhu, J. Y., Sabo, R., & Luo, X., (2011). Integrated production of nano-fibrillated cellulose and cellulosic biofuel (ethanol) by enzymatic fractionation of wood fibers. *Green Chemistry, 13*(5), 1339–1344.

INDEX

For Product Safety Concerns and Information please contact our EU
representative GPSR@taylorandfrancis.com
Taylor & Francis Verlag GmbH, Kaufingerstraße 24, 80331 München, Germany

9 781774 637517